THE *Spirit*

IN THE GENE

THE *Spirit*
IN THE GENE

COMSTOCK PUBLISHING ASSOCIATES *a division of*
Cornell University Press   ITHACA & LONDON

Humanity's Proud Illusion and the Laws of Nature

REG MORRISON

WITH A FOREWORD BY

LYNN MARGULIS

First published 1999 by Cornell University Press

Printed in the United States of America

Cornell University Press strives to use environmentally responsible suppliers and materials to the fullest extent possible in the publishing of its books. Such materials include vegetable-based, low-VOC inks and acid-free papers that are recycled, totally chlorine-free, or partly composed of nonwood fibers.

Librarians: A CIP catalog record for this book is available from the Library of Congress.

Cloth printing   10   9   8   7   6   5   4   3   2   1

CONTENTS

Foreword by Lynn Margulis   vii
Preface   xi
Acknowledgments   xvii

PART I: PROBLEMS
1. A Prattling Prodigy   3
2. Turn of the Tide   11

PART II: ORIGINS
3. Our Genetic Origins   57
4. The Agrarian Transition   91
5. Evolution's Answer to Biological Waste   102
6. Correcting Imbalances   119
7. The Terminators   144

PART III: SOLUTIONS
8. The Spirit in the Gene   159
9. Excalibur!   193
10. Midnight Prognosis   233

Notes   261
Bibliography   273
Index   281

FOREWORD

Reg Morrison's thesis is as intriguing as it is unique. Why do we convert our grasslands, woodlands, riverbanks, and even swamps to urban blightland? Why do we delight in the new house, the new clothes, the new music, and most especially the new baby at the expense of our lush coastlines and green meadows? Most germane, how did the hominids that evolved into modern humans triumph over their potential predators and leave their homeland to make habitable crannies on every continent, begetting today's milling crowd of 6,000 million people, Morrison's "plague mammal," his "prattling prodigy"?

Morrison's idea is that our dogged belief in human invincibility leads us to overpopulate the Earth, resulting in wanton habitat destruction throughout the glorious nonhuman world. I consider his analysis of the peculiar human mammal original yet highly defensible by standard science and other empirical modes of knowledge. Radical, defended and correct, this book is by a deeply interested author who has absolutely no vested academic, economical, egotistical or other usual kind of interest in his thesis. He has no axe to grind. Inspired by his prose but dissatisfied that the book's title simply did not reflect the power of its contents for months we (Reg, editor Peter Prescott and I) played with alternative suggestions. Among the most successful of our many attempts to title and subtitle this masterwork were the following five.

Habitat holocaust: Strong religious conviction and tribal loyalty insures maximum numbers of healthy children.

Origins of religion: Evidence that human spirituality, delusions of invincibility and ethnic superiority has been leading to environmental destruction.

Wounded Earth: Religious justification of agriculture-driven urbanization.

True believers: The genetic origin of religion, racism, and economic rationalization as basis of human population explosion and its unconscionable environmental consequences.

Excalibur: Anthropocentrism as evolutionary strategy.

Cornell University Press publishes, as do all good scientists and scholars, theses for which there is well-mustered evidence. The compelling argument in *The Spirit in the Gene* is based on a profound reading of scientific literature and on Morrison's personal experience; it is admittedly idiosyncratic. The prose, delightful and explicit, metaphorical and forceful, recreates Morrison's personal experience for the reader. The result, *The Spirit in the Gene,* cannot even be deeply criticized without well-developed counterevidence.

So what is this rare, fresh amalgam, written in the voice of an accomplished but relatively unknown writer? This, Reg Morrison's fourth book, is his first text-driven vehicle. With vision honed by his trade as a photojournalist, what follows reflects Morrison's observations and keen synthesis about nature: the air, land, and water system in which we are all inextricably and permanently embedded. I see his assertions in the context of the greatest sociologist of science, Ludwik Fleck (1896–1961), who in his masterwork makes clear that no bare scientific fact exists. (Those who seek detailed narrative can read Fleck's *Genesis and Development of a Scientific Fact,* 1979 English translation, University of Chicago Press, first published in Basel in 1935 as *Entshtehung und Entwicklung einer wissenschaftlichen Tatsache.*) Fleck notes that facts develop slowly and in a specific historical ambience. They represent and make known views of what he identifies as "the thought collective," those scientists who share a "thought style" or joint system of viewing the world.

New scientific knowledge makes its formal debut in any professional journal as statements tentative, peculiar, and often esoteric even to coauthors. The creative act of members of the thought collective is to organize, edit, and reshape the primary science accordingly. Fellow scientists imbue scientific pronouncements with authority and present them for members of their thought collective as textbook truth. But even a clearly stated new scientific fact is usually met with resistance by the collective, resistance that is eventually overcome and ultimately consolidated into an interlocking system of consistent ideas that, taken together, are comprehensible as part of a newly emerging thought style.

About scientific truth, Fleck says, "Very different thought styles are used for one and the same problem more often than they are used for very closely related ones. For example, it happens more frequently that a physician simultaneously pursues study of a disease from a clinical-medical or bacteriological viewpoint together with that of the history of civilization, than from a clinical-medical viewpoint together with a purely chemical one." Fleck says that it is, for instance, possible to trace the idea of an infectious disease from a primitive belief in demons through the idea of a disease miasma to the theory of the pathogenic agent. "Truth is not 'relative' and certainly not 'subjective' in the popular sense of the word. It is always, or almost always, determined within a thought style. One can never say that the same thought is true for A and false for B. For if they [people A and B] belong to the same thought collective, the thought will be either true or false for both. But if they belong to different thought collectives, it will just not be the same thought! It must either be unclear to, or be understood differently by, one of them. Truth is not a convention, *but rather (1) in historical perspective, an event in the history of thought, (2) in its contemporary context, stylized thought constraint*" (italics in original) (Fleck p. 100).

When he wrote about the "tenacity of closed systems of opinion," Fleck made clear that members of a widely held thought collective participate in a kind of "harmony of illusions." A keen observer, Morrison talks directly to this issue from his own guts and heart. He criticizes today's "harmony of illusions." The illusion that most concerns him is the one that leads us *Homo sapiens,* after all only one species of animal in today's world of more than 7 million, to spoil our own nest as well as the nesting sites of so many of our planetmates. I have no doubt that some readers, even erudite ones, will turn apoplectic, screaming that Morrison has his facts wrong. But Fleck was wise in his understanding of how new facts are received by a thought collective, and Morrison must be seen in the context of the new thought style he is helping to generate. Recall, Fleck noted, that scientific facts begin with resistance by the collective. Only after this initial reluctance are the facts consolidated into interlocking, consistent belief systems. Since in this sense the facts don't yet exist as such, Morrison cannot be wrong about them; he can only observe and describe our planetary turmoil from the scientific thought style this book begins to create.

Morrison peruses the stunning Australian landscape with a photographer's eye and a philosopher's thoughtful exertion. He is developing an international thought style based on the writings of those who belong to various scientific thought collectives. Morrison offers us a new, scientific look at what I see as the origin and evolution of religion. He calls this mysticism, but I see it as religion in the broadest sense: that which binds people together.

Using coherent narrative throughout, chock-full of observations and commentary, facts and interpretations, Morrison walks us through his plausible explanation for human origins and evolution. For many years he has studied Australia's outback—the banded iron formation, the microfossil, the desert, and the retiarius spider. Not a scientist nor really a scholar, only a voracious reader who thinks for himself, our author belongs to no scientific thought collective. Population ecologists, primatologists, behaviorists, and sociobiologists all view one and the same problem in different ways. Belonging formally to none of these collectives, Morrison sees the problem of humankind's devastating impact in his own way.

Morrison most often quietly, but sometimes with a booming voice, leads us into the song-lined bush, where he bids us watch, listen, and think a moment with our big brains and not with our loins. Some of his trails are well trod; others meander or detour. Many are newly blazed. All are exhausting and exhilarating as they beckon us to follow. The perspective from the summit is well worth the trek as we are goaded onward to his lookout from where we can peruse the limited horizon. From the vantage point of our evolutionary history as teachable mammals and upright African primates, Morrison helps us to broaden the view of our one and only living planet and come to see our lives as they really are for the first time.

LYNN MARGULIS

Distinguished University Professor,
Department of Geosciences
The University of Massachusetts at Amherst,
February 1999

PREFACE

Aword of warning to those who venture past this page. The road ahead is somewhat tortuous, but it mirrors my own mental journey over the past two decades. It was a journey that led me eventually to a world where logic stands on its head and mere molecules make puppets of us all—a factual world that makes the wildest figments of imagination seem pedestrian by comparison.

My journey began with an attempt to define for my own peace of mind the full extent of the damage wrought by the present catastrophic explosion of earth's human population. The graph of human population growth over the past ten thousand years is disturbingly similar to the population graph of an animal entering what we would commonly describe as a plague phase. Similarly, the dismal litany of environmental damage that has occurred in this time makes the word *plague* unavoidable. But the term *plague animal* makes us think of rats, mice, and rabbits: surely it is absurd to link *Homo sapiens,* that uniquely sentient and rational offshoot from the ape family with such pests? How can our painful ascent to civilization constitute, in any sense, a plague? This enigma led me to search in turn for the origins of the behavior that produced both this global phenomenon and our coincidental rejection of our genetic heritage as normal animals—animals that might even be considered maladapted and incompetent by animal standards. And this led inevitably to a search for the source of our species' astonishing evolutionary success in the face of overwhelming odds. For example, when the onset of the most recent ice age left our forest-bred ancestors stranded on the drying plains of East Africa 2 million years ago, why did not the leopards pick them off, one by one? Was their survival due simply to the development of language, technology, and a sharper mind? Perhaps. But the more one examines our talent for rational thought, the less one can ignore our peculiar capacity for self-delusion and self-destruction that seems to go hand in hand with it. If these traits also evolved on those dangerous African plains, why did not these inherent flaws outweigh the potential advantage of the rational brain and tip the scales in favor of the leopards once again?

The more one examines behavior for clues to its evolutionary origins, the more one becomes entangled in the underpinning chemistry and neurobiology; inevitably, the deeper I delved into those two jargon-ridden jungles the more often I came upon paths that led directly to the genes. But if the biological evidence is to be believed and DNA is indeed the underlying source of all animal behavior, including human, then two other crucial questions are raised: (1) How are human genetic imperatives translated into neuronal activity and incorporated into our behavior without the rational brain becoming aware that it has been bypassed? (2) If genes are the source of all behavior, what part, if any, does culture play?

Charles Darwin's proposition that we humans evolved from an ancestor of the ape family now offends only religious fundamentalists and the uneducated. And yet the obvious corollary that humans are consequently typical primates and entirely driven by their peculiar genetic heritage seems to offend almost everybody. People accept that the genes are the sole driving force for animals because animals are just that, mere animals. By contrast, they say, we humans are unique in that we possess both a spiritual and intellectual capacity separate from and independent of our physical body. Consequently there is consensus across all cultures that this independence of mind bestows both a unique sense of self-awareness and the ability, even responsibility, to tailor our behavior to fit the dictates of reason and to countermand what are considered the baser instincts of our animal heritage.

The reader should be aware that in order to emphasize a major theme of this book—that we humans are solely the product of evolution and therefore entirely typical mammals—I reiterate a point made by Jared Diamond, Carl Sagan, and others, that the classification of humans as

Hominidae, a family distinct from the great apes, is clearly untenable in the light of comparative DNA studies. In fact, the difference between the active portion of human and chimpanzee DNA is so small that we should even place ourselves within the chimpanzee genus. In this text, therefore, I occasionally refer to humans not only as one of the great apes but, in the words of Jared Diamond, the third chimpanzee.

The curious perception prevalent among humans that unlike other creatures each of us exists as two separate entities—one physical, the other spiritual—appears to have attended our species throughout its recorded history. It has been the Gordian knot of all philosophy from Aristotle onward and continues to provide humanity with most of its triumphs and tragedies and all of its moral dilemmas. In other words, this dual image of ourselves is the source of both our pleasure and our grief. Spirituality shapes our feelings of affection, admiration, compassion, joy, and hope and weaves into our culture the glittering threads of passion that enliven our literature, music, dance, drama, and art. This dichotomy of body and spirit also represents the source of most human misery, causing us to succumb to the general belief that we are morally accountable for our misdeeds.

Most of us expect both moral and rational perfection from ourselves and others. Human frailty irritates us. Meanwhile the brain is often spoken of as though it were an architect-designed, fully integrated unit, rather like a computer—but one generally driven by novices and idiots. The human brain in reality is more like an old farmhouse, a crude patchwork of lean-tos and other extensions that conceal entirely the ancient amphibian-reptilian toolshed at its core. That it works at all should be cause for wonderment. As for pointing to our mental failures with scorn or dismay, we might as well profess disappointment with the mechanics of gravity or the laws of thermodynamics. In other words, the degree of disillusionment we feel in response to any particular human behavior is the precise measure of our ignorance of its evolutionary and genetic origins.

In the introduction to his book *The Blind Watchmaker,* the distinguished British evolutionist Richard Dawkins was even moved to complain: "It is almost as if the human brain were specifically designed to misunderstand Darwinism, and find it hard to believe."[1] Our universal, and therefore genetic, need to see ourselves as separate from the rest of the animal world ensures that most of humanity will continue to be at least suspicious, if not thoroughly antagonistic, to Charles Darwin's heretical propositions. We conveniently contend that we alone of all earth's species are not normal animals, an extraordinary claim that demands extraordinary proof. And none exists.

Not the slightest scrap of hard evidence, either morphological or genetic, exists to suggest that *Homo sapiens* is not, like all other animals, a natural product of evolution. Therefore we, like they, are uncontaminated by supernatural influences, good, bad, or divine. We may well be excellent communicators and toolmakers, and the most logical, self-aware, mystical, and malicious animals on earth, but overwhelming evidence shows that these distinctions are of degree, not of kind. The only irrefutable argument in favor of humanity's specialness is in fact purely mystical—and entirely circular. Yet the myth lives on.

Is it not strange that our genetic makeup should allow, perhaps even prescribe, such naïveté? I will argue that our peculiar genetic heritage purposefully blinds us to reality to make us malleable and compliant to its demands, and that our habit of assigning ourselves an imaginary specialness is the mechanism that delivers us willingly into genetic servitude. Our purported spirituality is a consequence of 2 million years of painstaking Darwinian selection.

Having evolved as a cooperative species, *Homo sapiens* seems to have retained almost all of those mammalian characteristics we most admire—selfless devotion, compassion, courage, generosity, and wit—to the point that one of the truly remarkable things about human beings is not how bad we can be but how good most of us are, most of the time—even by animal standards. Unfortunately we take this goodness for granted and usually fail to note it. When we do, however, we assume it to be uniquely human, an expression of human spirituality. In fact, altruistic behavior is common throughout the animal world, and in other species it seems to be entirely free of tedious sermonizing and self-congratulation. They, like us, simply do what works best for their genetic line. To this end, many species mate for life; feed, protect, and educate their offspring with obsessive fervor; and willingly lay their lives on the line whenever family, tribe, or territory is threatened.

The attribution of human motives and emotions to animals used to be considered sloppy science. The underlying fear was that such thinking might erode some of the respect that we felt we were owed as a uniquely sentient and rational species. That particular academic taboo is less rigidly observed these days, yet in a perverse sense it remains entirely sound. Indeed, no animal displays human behavior. Quite the reverse. Humans display only animal behavior. Watch the action without the sound track and this truth becomes obvious.

I will therefore argue that our much-vaunted spirituality is a cultural illusion that became cemented into the foundations of early human society by our potent combination of language and imagination. Meanwhile the universality of our fascination with things mystic and spiritual displays its genetic origins as plainly as does our compulsion to communicate with one another. I also believe that our obsessive urge to imbue our existence with mystical meaning was once the Excalibur[2] of our species, the invincible weapon that carried our branch of the hominid line from the brink of extinction to the conquest of the planet. Since mystical beliefs of various kinds have also played a primary role in the catastrophic growth of the human population, the final chapters of the book are devoted to exploring mysticism's present and future impact on our already bruised and destabilized environment.

We may not be able to hurl our troublesome Excalibur back into the gene pool from whence it came, but surely it is time we momentarily lowered that dazzling weapon, and for once, with unclouded eyes, saw ourselves for what we are in the only context that ultimately matters: the evolutionary context.

ACKNOWLEDGMENTS

Writers who venture into the technical minefields and tribal territories of science without good local guides tend to have a short life expectancy. My guides were few, but they were the best, and I thank them most fervently for their generosity and dedication.

I cannot adequately express my gratitude to microbiologist Lynn Margulis. Her painstaking technical editing and general guidance, especially in matters bacterial and evolutionary, added a vital new dimension to the book, meanwhile her support and encouragement were invaluable to me personally. I am similarly indebted to neurobiologist David Sandeman of the University of New South Wales for his unflagging assistance, patience, and encouragement. Others who took pains to guide me through the minefields were geologist Malcolm Walter and geneticist Keith Williams of Macquarie University, archaeologist Alan Thorne of the Australian National University, ecologist Chris Dickman of the Sydney University, and Tim Flannery, zoologist and senior research scientist with the Australian Museum. A special thank-you also to Mary White, old friend, author, paleobotanist, and passionate conservationist, for her continued support and invaluable advice. And finally, an inadequate thank-you to my patient, lion-hearted wife, Pauleen, for the endless midnight copyediting, and all the other dreary chores that enable books such as this to reach the publisher.

PART I

Problems

Chapter I

A Prattling Prodigy

Words, words, words . . .
—*Hamlet,* act II, scene ii

There is a slight bulge on the left side of my brain that has a lot to answer for. Not only does it regularly make an ass of me by causing me to talk aloud to myself, but it is also responsible for some rather more worrying phenomena, such as Africa's famines, both world wars, the AIDS epidemic, and even global warming! I hasten to add that this is not an expression of neurotic guilt on my part, for my bulge is no mere personal defect. It is, instead, a kind of neuronal heirloom that has been in my family—and yours— ever since the first humans roamed the drying plains of East Africa some 2.5 million years ago. Indeed, that small bulge is probably the only credential that really matters in our claim to membership in the species *Homo sapiens.*

Viewed in this light, that little emblem of humanity suddenly assumes monumental proportions. In the last 10,000 years alone, we humans, acting under its spell, have succeeded in reworking most of the face of this planet and have utterly changed the course of all future evolution. Yet without that small bulge you and I would be grunting cave dwellers. The secret of its power? "Words, words, words": that faint bulge is the midwife of all human language.

We generally prefer to credit humanity's astonishing array of cultural and technological achievements to three other characteristics: our peculiar manual dexterity, our capacity for deductive reasoning, and our unbridled imagination. But it was the development of language that first gave imagination its wings and then allowed our ancestors to preserve and pass on each hard-won cultural and technological advance so that succeeding generations might benefit from them. Language provides the glue that holds society together—and on occasion manufactures the bombshells that tear it apart. Words even have a life of their own at times, appearing in the brain unbidden by any conscious thought process and rolling off the tongue despite our better judgment. And although language is used primarily for communication between people, the lack of a listener is no impediment; most of us will mutter happily to ourselves. Until the age of seven, children talk to themselves 20% to 60% of the time. According to research in both Russia and the United States, talking to oneself is not only normal but necessary for healthy mental development.[1]

We humans are compulsive communicators, with brains that seem to be hardwired with a few basic rules of grammar. These rules are applicable to all forms of language, whether spoken, written, or sign language. We know roughly where in the brain this complex process occurs, but we don't yet know how. The region of the brain that puts it all together for us is an ill-defined, neuron-packed swelling that lies just above and behind the left eye. To those who specialize in things cerebral, this language factory is known as Broca's area, after the French surgeon and pioneer anthropologist Paul Broca, who first identified its peculiar function in 1861.

Broca's Language Factory

While many other animals have a generalized communication area within the brain, no other has a highly specialized language center like ours. Even our own evolutionary sibling, the chimpanzee, although equipped with a generally similar brain structure and a communication center localized in the left frontal lobe, shows little sign of Broca's area.

Our Evolutionary Siblings

Chimpanzee brains are considerably smaller than ours, with about a quarter of the surface area of a human brain (see figure 1). They consequently

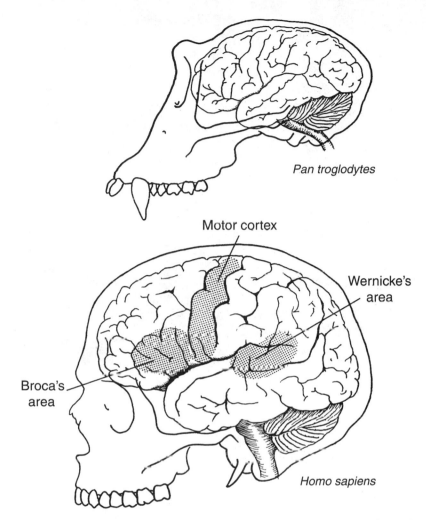

Figure 1. Human and chimpanzee skulls, shown here in their true size relationship.

possess much less gray matter, or cerebral cortex, where most thinking occurs. Such measurements were once regularly quoted as scientific evidence not only for our superior intelligence but also for the proposition that we had been either separately created or divinely set apart from the rest of the animal world. It was accepted that of all the world's species we alone were logical, imaginative, and sentient. The misconception lingers on. More rigorous and objective research shows that brain size is a very unreliable guide to intelligence. Many other animals are capable of rudimentary logic and relatively complex communication. Even more startling is the discovery that chimps, gorillas, and orangutans are capable of using

Common chimpanzee (*Pan troglodytes*). An old male chimpanzee submits to some painstaking facial grooming from one of the troop's senior females.

formalized sign language (since speech is anatomically unavailable to them) to express a limited range of imaginative desires and feelings to their human companions.

The revelation of these disturbing streaks of humanity in the other great apes generated academic controversy around the world during the 1970s and 1980s. It should not have. Genetically we are barely distinguishable from chimpanzees. Rather it is the marked physical difference between our species that should surprise us. It is often said that we humans share about 98.4% of our total DNA with chimpanzees. But at least 95% of this is "nonsense" DNA, vast stretches of repetitive gobbledygook that remains unexpressed in either animal. More illuminating is the comparison between the active components of DNA, the genes. These protein-coding sequences show a 99.6% match.[2] Of the two extant species of chimp, the bonobo, or pygmy chimpanzee (*Pan paniscus*), appears both morphologically and behaviorally to be most closely related to humans. The tiny 0.4% genetic difference guarantees us a seat beside them on the primate family tree. Despite a superficial similarity, chimps and gorillas are mere cousins by comparison, sharing only 98.2% of their total DNA. Orangutans, on the other hand, branched off on their own some 15 million years ago and, sharing only 96.7% of their DNA, are distant cousins.

Judging by the molecular evidence, therefore, our generic classification is unequivocal: chimpanzee. Either we should be cataloged as *Pan sapiens,* the third chimpanzee, or more properly all three species of chimp ought to be reclassified under our present generic label, *Homo,* the term originally assigned to chimpanzees by the founder of modern taxonomy, Carolus Linnaeus. Linnaeus subsequently invented a taxonomic distinction between humans (Hominidae) and chimpanzees (Pongidae) at the family level, but only out of a well-founded fear that if he didn't he would outrage not only the Swedish Lutheran Church but the whole of society, religious and secular.[3]

The inescapable corollary of the close relationship between chimps and humans is that shared characteristics represent genetic factors inherited from a common ancestor. The streak of humanity that we see in each of the other great apes is, in reality, a suite of ancestral characteristics now shared by all of its descendants, including us.

Anthropocentric Myths

Until Charles Darwin published *The Origin of Species* in 1859, most people had been able to ignore these obvious but embarrassing family connections. Subsequent research has now amassed such overwhelming physical, chemical, and genetic evidence of the evolutionary link between the species that Darwin's original detractors now seem absurd. Nevertheless, among well-educated people, even among those who freely concede this evolutionary link, the assumption remains that there is a vast gulf between humans and apes. Bedazzled by the ancient myth of anthropocentrism,[4] many still fail to comprehend that the evolutionary gulf that appears to separate human beings from the other great apes is largely a cultural construct, with profound consequences, that has little to do with fact.

Many outward differences distinguish *Pan* from *Homo:* our upright carriage, facial structure, and relative lack of body hair. But since all breeds of dogs are, by genetic definition, members of the same species, appearance is an untrustworthy measure of species distinction. Is behavior a better guide?

Predominant among the behavioral features that separate us from our chimpish relatives is our facility with language. The urge to communicate and at least some of the mental structures that enable language are common to all the primates and to the apes especially. Our peculiar ability to comprehend and maintain grammatical structure, the trait that allows us to distinguish instantly the crucial difference between "Can you see the leopard?" and "The leopard can see you," is limited to humans. Grammatically structured communication enables us to string together a related sequence of thoughts like beads on a necklace. Change the order of the verbal beads and you may dramatically change the meaning.

This peculiar talent resides first in the brain's language production center, Broca's area, and second in an adjacent decoding department known as Wernicke's area. The development of these linked communication cen-

Bonobo (*Pan paniscus*). Slimmer, more graceful and less aggressive than chimpanzees, bonobos may well represent the genetic center line from which both chimpanzees and humans diverged.

ters in close proximity to the neighboring motor control area seems to have underwritten our astonishing evolutionary success. Communication by commonly understood sounds is certainly not unique to us, but grammar allows us to communicate even complex information with passable precision. All the other great apes—gorillas, orangutans, and chimpanzees—can learn the rudiments of human language, yet none have achieved more than the simplest kind of communication. If Kanzi, a sharp-witted bonobo, wished to express contrition or sadness via the computer-generated symbols taught to him by primatologist Sue Savage-Rumbaugh, he could make the sign for "Kanzi" followed by the sign for "sorry." Had he wished to elaborate, however, his only option would have been to repeat the two signs.[5] We have no such limitations.

> O! that this too too solid flesh would melt,
> Thaw, and resolve itself into a dew;
> Or that the Everlasting had not fix'd
> His canon 'gainst self-slaughter! O God! O God!
> How weary, stale, flat, and unprofitable
> Seem to me all the uses of this world.
> Fie on't! O fie! (*Hamlet*, act I, scene ii)

Precisely what it is in genetic or neuronal terms that allows human beings to express anguish on this scale and at this level of complexity we may never fully comprehend, but we should look back in awe, for here lies the gulf that most effectively separates our species from all others.

It was the flexibility of grammatically structured language that enabled our forebears to cooperate more effectively in hunting and gathering their food, to divide and share it more readily, and to teach their children to do the same. Similarly, precision of speech enabled them to disseminate and store within their tribes vital environmental information, hunting skills, and craft techniques. More important, it enabled tribal elders to pass this accumulating reservoir of knowledge to their children, giving them a head start in the struggle for survival. By such means, these talking primates not only clawed their way to the top of the food chain but eventually spread far beyond tropical Africa, even to the most inhospitable regions of the planet.

Earth's Prattling Prodigy

We would seem to be without peer: earth's prattling prodigy. Yet as we approach the end of the twentieth century, a century in which we consolidated our communications around the planet, ventured into space, and even placed a shaky hand on the molecular levers of life itself, ominous signs emerge of a massive global backlash. No longer confined to a few trouble spots, the indications of environmental disturbance now extend to the fringes of the biosphere.

Today's scenario engenders bitter confrontations between conservationists and their detractors, and acrimonious debates rage around the world. Traditional villains, both individual and corporate, are denounced almost daily in the media. But are there villains to blame, or are more complex forces at work? If our growing environmental problems are truly attributable to human activity, do certain individuals or groups deserve our special condemnation, or have we all behaved badly? Is it possible that this successful species of ours also embodies one or two heritable flaws against which we have no defense, flaws embedded in the wiring of our brains, or hidden in the coils of our massive DNA?

Answers to these questions will provide the keys to both our past and our future, and in some respects, the future course of evolution on this planet.

Chapter 2

Turn of the Tide

Where is the thicket? Gone.
Where is the eagle? Gone.
The end of living and
the beginning of
survival.[1]

Human Expectations

No longer is any secret place on earth without a trace of human impact. Carried around the globe by wind and water, the chemistry of our existence now affects almost all other life. Meanwhile, as the human population continues to explode and technology expands, our impact on this planet multiplies.

Like all land mammals, we depend on four primary resources for our daily survival: breathable air, potable water, arable soil, and a diverse, supportive biota. The energy we humans daily appropriate, directly or indirectly, to feed, clothe, and house ourselves amounts to somewhere between 20% and 40% of all the solar energy that is photosynthetically trapped each day by vegetation on the world's landmasses (see figure 2).[2] We then supplement this incoming solar energy by drawing upon previous solar income that was harvested by life in the past and is now stored within the earth's crust in the form of coal, oil, and gas. Unfortunately the 300 million metric tons of humanity that the earth currently supports has an appetite

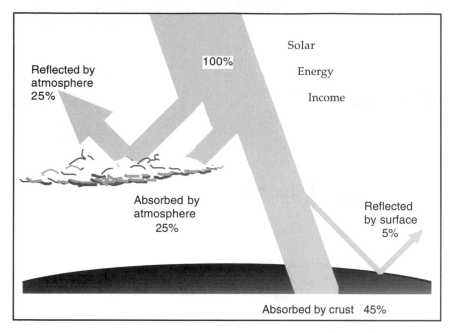

Figure 2. Global energy income.

so voracious that the planet and its biota can meet our demands only by divesting itself of vast numbers of other energy consumers.[3] Our survival thus depends on the wholesale extinction of other species. We live by committing bioticide.

The problem stems largely from the human habit of expecting much more from life than other animals do. Even worse, we tend to view our high level of consumption as perfectly acceptable, even admirable. Consequently those of us fortunate enough to be born to average-income parents in developed countries come into this world bearing an astonishingly high environmental price tag for an animal our size. This level of consumption by humanity is not a constant of course, since it is governed entirely by fluctuations in the size of the global population and the level of technology employed. One average American, Australian, or western European annually consumes as much commercial energy as 120 average inhabitants of a country such as Bangladesh.[4] Moreover, the extraction of energy and materials by inhabitants of wealthy nations is not confined to their own region but is spread around the planet through globalized industry and trade. The orange juice I drank for my breakfast in Australia may have contained imported pulp and contributed to the burning of a rain forest in Brazil. In achieving this degree of dominance over the world's natural ecosystems, humans during the past 10,000

Brush-tailed possum (*Trichosurus vulpecula*). Lost, and dazzled by morning sunlight, this brush-tailed possum epitomizes the plight of wildlife displaced by urban expansion all around the world.

years have transformed both the face of the planet and the atmosphere that envelops it.

Like all of the 4.5 billion years that have preceded it, the current year will be measured out by the daily spin of the earth as it shoulders its way through the icy dust of its solar orbit. The seasons will follow in their customary order, regenerating a tide of life that first began to surge about this planet at least 4 billion years ago. Once more it will seem that order and stability are inherent in the biosphere. Sadly (from our anthropocentric point of view) this is simply not the case.

Scottsdale, Tasmania. If an insect had done the damage shown here, no expense would be spared to eradicate the pest. But the devastation in this case was legally inflicted by loggers harvesting virgin Tasmanian rain forest for sale to Japan as wood-chips.

The fossil record tells us that the fabric of life has been ripped to shreds at least five times in the past 500 million years by massive global disturbances. And now it seems this ancient cycle is about to be repeated. Species are disappearing at a rate comparable to the period following the catastrophic events of 65 million years ago, when a fragmenting comet slammed into the planet. The vast repercussions of its impact killed off the last of the world's dinosaurs and removed some 65% to 70% of all known species.[5] Meanwhile we are besieged by a host of environmental problems—land degradation, loss of forests, air and water pollution, a decaying ozone layer, and accumulating evidence that global weather patterns are becoming increasingly unstable.

So serious are these problems that Harvard professor and two-time Pulitzer Prize–winner Edward O. Wilson, one of the world's most respected biologists, believes that we will enter the twenty-first century with somewhere between 10% and 20% of earth's prehistoric inventory of species extinguished and one of its most successful organisms, *H. sapiens,* penciled in under the heading "threatened."[6]

The main threat appears to be the general erosion of our natural resource base. The air is less breathable, the water less drinkable, and the grain resources on which humanity depends are shrinking yearly on a per capita basis. Meanwhile the species that support our existence are fewer and more vulnerable than ever before, and those pathogenic species that feed on them—and on us—multiply and grow stronger every day. When superimposed on the general decline in biodiversity around the globe, these mounting problems represent the symptoms of severe habitat decline, and as such, portend catastrophe for an environmentally expensive animal like ourselves.

Assault on Diversity

The primary measure of biospheric stress is loss of species. For life to evolve and flourish, new species must, on average, evolve faster than old species become extinct. Apart from a number of spectacular but relatively brief interruptions, such has been the case for most of the 4 billion years during which there has been life on this planet. According to the fossil record, the past 65 million years have been the most genetically fruitful of all, but that trend has now reversed. Edward O. Wilson laid out the problem as follows:

> Even with . . . cautious parameters, selected in a biased manner to draw a maximally optimistic conclusion, the number of [rain forest] species doomed each year is 27,000. Each day it is 74, and each hour 3.
>
> If past species have lived on the order of a million years in the absence of human interference, a common figure for some groups documented in the fossil record, it follows that the normal "background" extinction rate is about one species per million species each year. Human activity has increased extinction between 1,000 and 10,000 times over this level in the rain forest by reduction in area alone. Clearly we are in the midst of one of the great extinction spasms of geological history.[7]

Tropical forests are the last great bastion of biodiversity on this planet, providing habitat for almost half the world's land-based plant and animal species. Having shrunk to half their prehistoric area, they currently occupy only about 6% of the earth's surface and are disappearing faster than ever. According to satellite data assessed in the early 1990s by the United Nations Food and Agriculture Organization (FAO) and by the Washington-based World Resources Institute, the planet was losing the last of its forest cover at the rate of about 207,000 square kilometers

(80,000 square miles) a year. That rate of loss was 40% to 50% greater than it had been a decade earlier, and the rate has continued to accelerate throughout the 1990s. The world is entering the twenty-first century with little more than 10% of its original forest cover intact, and by 2050 this too will have largely disappeared, according to paleoanthropologist Richard Leakey and Roger Lewin in their book *The Sixth Extinction: Biodiversity and Its Survival.*[8]

Aided by unprecedented drought in 1997, deliberately lit fires raged for many months in most of the world's tropical forests. Normally free of fire, tropical forests have few mechanisms for regeneration and may be permanently destroyed by burning—along with the thousands of species that are endemic to them. A World Wide Fund for Nature (WWF) report claimed that more tropical forest burned around the world in 1997 than at any other time in recorded history. In the words of WWF Director General Claude Martin, "This is not just an emergency; it is a planetary disaster."[9] The following year, when the drought resumed, many of the old fires reignited from unquenched embers in smoldering peat bogs, and hundreds of new fires were lit, as usual, by loggers and subsistence farmers.

Most forests are felled or burned to provide timber, to create farmland, and to make room for urban expansion. Such processes guarantee the demise of many thousands of species every year. Globally, the extinction rate may be more than 30,000 species a year (about 3.4 an hour) according to Leakey and Lewin. However, estimates range from 17,000 to 100,000 extinctions a year. Biologists have now documented about 1.75 million living species, yet they have little idea what percentage of the biota that figure represents. Most authorities estimate that the biota contains between 10 million and 100 million species. With that uncertainty in mind, the 50% species loss that Wilson and others predict by the year 2100 sounds like an irresponsible guess. On the other hand, we can be fairly certain that the percentage of extinctions that we don't see is similar to the percentage of species we don't know about, which leaves extinction percentage estimates intact, whatever the true size of the biota. Given a background extinction rate of only one species per million per year, an extinction rate some 30,000 to 100,000 times that size represents a biological catastrophe that must soon threaten even *Homo sapiens*.

Past Extinctions

Viewed against the larger scale of geological time, the present episode of mass extinctions is nothing new. Others have occurred in the past, long before the evolution of human beings. The fossil record of the past 500 million years clearly shows that similar episodes have occurred on many occasions. Five of these stand out in particular (see figure 3). They are

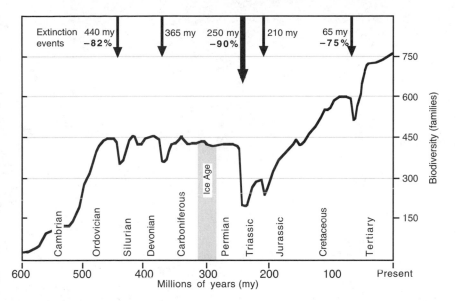

Figure 3. Mass extinctions. The arrows indicate the five major extinction events of the past 600 million years. The percentage losses shown against three of the events are the estimated species extinctions.

listed here according to the periods they terminated, together with the approximate time they occurred: Ordovician (440 million years ago), Devonian (365 million years ago), Permian (250 million years ago), Triassic (210 million years ago), and Cretaceous (65 million years ago).

Each of these extinction events appears to have been accompanied, if not caused, by major disturbances in the global climate, and most, if not all, may have been triggered by cosmic impacts.[10] Two of these extinctions, in the Permian and Cretaceous periods, have particular significance for us. The Cretaceous event was crucial in that it removed dinosaurs from the evolutionary stage, allowing mammals, and ultimately ourselves, to achieve dominance. The Permian event, although ill defined, is nevertheless significant because of its sheer size. According to Chicago paleontologist David M. Raup and Washington paleobotanist Douglas H. Erwin, detailed analyses of the Permian fossil record suggest that the global biota had an extremely close brush with total destruction at this time. Best estimates suggest that roughly 50% of families and 70% of genera were wiped out, and this known casualty list suggests the extinction of somewhere between 80% and 95% of all species.[11] The dimension and duration of these catastrophes suggest that once a mass extinction has begun, it not only ripples through the contemporary biota but also echoes on through time in a chain reaction so complex and far reaching that its passage and limits lie well beyond the scope of human comprehension.

Gosse Bluff, central Australia. This circle of hills, known as Gosse Bluff, is the remains of a deep crustal bruise left in central Australia by a small comet about 130 million years ago. The original crater would have been about 22 kilometers (13 miles) across—a pinprick compared to the cosmic impact that was to bring the biological abundance of the Cretaceous period to a cataclysmic close some 65 million years later.

As the fossil record of these extinctions clearly shows, global biodiversity is a house of cards. With this in mind our current assault on it appears to verge on the suicidal. In the words of Edward O. Wilson, "every species extinction diminishes humanity."[12] It is not, however, the demise of evolutionary masterpieces like the snow leopard, the mountain gorilla, or Spix's macaw that will undo us, but rather the unnoted hemorrhage of the biota's minute workhorses—everything from bacteria to beetles—that will ultimately prize this planet from our grasp.

Global Warming

Without earth's thin blanket of atmosphere, and without the organic chemistry that maintains it, this planet would be a lifeless ball of frozen rock, approximately 33°C colder than at present. The average temperature

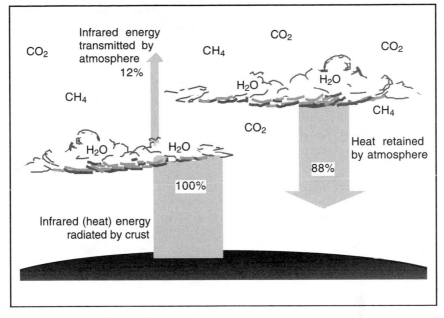

Figure 4. The greenhouse effect.

on the moon, which is airless, is about $-18°C$. The average temperature of the air just above the earth's crust is around $15°C$.

About 25% of the solar energy that reaches the planet is reflected back into space by ice crystals, dust, water vapor, and other aerosols in the atmosphere. Another 25% is absorbed by the atmosphere as it passes through, and 50% reaches the earth's crust. The crust then radiates some of this energy back into the atmosphere at infrared wavelengths. Of this radiant heat about 88% is then absorbed by atmospheric gases, allowing only 12% to escape back into space. This is known somewhat inaccurately as the greenhouse effect (see figure 4). This process, which makes earth habitable, is due to the presence of several heat-absorbent gases, notably water vapor, carbon dioxide, and methane.

At a molecular level, the principal heat sinks are the oxygen and carbon atoms that characterize the structure of the greenhouse gases. They consequently constitute the primary thermostat that controls the surface temperature of the planet and determines its human carrying capacity.

During the past two hundred years of accelerating industrialization, humans have burned vast quantities of fossil fuels such as coal, oil, and gas in order to provide heat and power. Since each atom of this fossil carbon combines with two atoms of oxygen during the burning process, each metric ton of carbon burned adds 3.7 metric tons of carbon dioxide to the atmosphere. This massive liberation of biologically sequestered carbon has boosted the atmosphere's carbon dioxide component by about 30%, thereby significantly increasing its capacity to retain heat. From 1850 until 1950 about 60 billion metric tons of carbon were burned. We now inject that amount of carbon into the atmosphere every seven or eight years, adding more than 6 billion metric tons annually by burning fossil fuels. In the same amount of time another 2.5 billion metric tons of carbon is added to the atmosphere as a result of deforestation.

The fires that ravaged the drought-stricken forests of South America and Southeast Asia in 1997 and 1998 provided another major injection of carbon into the atmosphere. The huge peat bogs of Sumatra and Borneo contain up to one hundred times more carbon per acre than the surrounding forest, and where a forest fire transfers to the peat it becomes unquenchable except by monsoonal flooding. However, in 1997 and 1998 the normal monsoonal rains failed to arrive due to a temperature-driven climatic disturbance commonly known as El Niño.

El Niño and the Southern Oscillation

El Niño (a reference to the Christ Child) was the nickname originally applied by Spanish-speaking South American fishermen to an eastward flow of warm surface water that frequently occurred in the eastern Pacific around Christmastime. During hotter years, however, this temperature-related current reversal can become so powerful that it affects ocean currents and weather patterns around the globe. On such occasions the world's equatorial trade winds may cease entirely, causing a failure of the monsoonal rains that water most of the earth's tropical regions. In the Pacific region the atmospheric component of this life-threatening phenomenon is known as a Southern Oscillation, and the combined pattern of El Niño and the Southern Oscillation is commonly referred to by its acronym, ENSO (see figure 5).

Major ENSO events typically produce violent storms along the U.S. Pacific coast, savage droughts in Africa and Australia, and a failure of the monsoon in Asia. Each ENSO cycle is then followed by a climatic backlash that results in more violent storms and widespread flooding, but this time in the drought-ravaged areas. Recent ENSO events have seriously disrupted human food production and cost tens of thousands of lives through famines, fires, floods, and diseases they have engendered.

Such major climatic perturbations used to occur only once or twice a decade. Since 1970, however, the mean annual air temperature in the eastern Pacific has risen by about 0.5°C, triggering a series of ENSO events that not only have been more frequent and more savage but have also

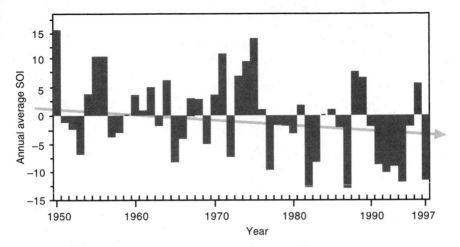

Figure 5. Southern oscillation index. The southern oscillation index (SOI) measures fluctuations in the barometric imbalance between air over the east Asian and east Pacific regions. Sustained high temperatures and low pressures over the southeastern Pacific generates a negative index, and in extreme circumstances (below SOI −5), produces the pattern of ocean currents known as El Niño. (*Source:* National Climate Center, Melbourne, Australia.)

tended to last much longer. The four-year Southern Oscillation pattern that ended in June 1995 was unprecedented in a hundred years of barometric data,[13] and there now seems little doubt that the increasing frequency and intensity of ENSO events is symptomatic of global warming. The evidence at a molecular level, however, is unequivocal.

Bubbles in the Ice

Drill cores 3 kilometers (1.8 miles) long, made at the Vostok research station in Antarctica, provide us with a cross section of the polar ice caps and a record of local precipitation that spans almost 300,000 years. Of special interest are the air bubbles that became trapped in the layers of ice when they were originally deposited. All air contains two isotopes of oxygen, each possessing a slightly different nuclear structure; and the ratio between these isotopes (oxygen-16 and oxygen-18) varies according to the ambient air temperature. Since polar ice accumulates seasonally in definable layers, these layers may then be read chronologically, like tree rings. By analyzing the air in the bubbles and matching the fluctuating levels of carbon dioxide and methane against the sequence of isotopic oxygen ratios, researchers found that they had an accurate polar temperature chart covering the time spanned by the layers of ice in the drill cores. All the ice-core records tallied precisely for the past 113,000 years, and most

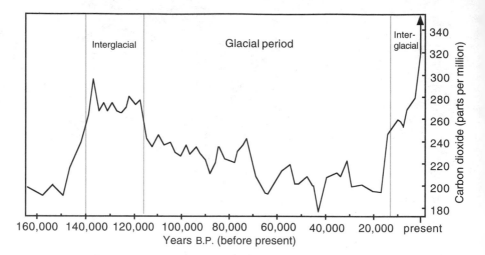

Figure 6. Atmospheric carbon dioxide. (Based on Vostok ice-core data.)

could be read with reasonable accuracy—and coincided tolerably well—for almost 250,000 years. These records clearly indicate that levels of atmospheric carbon and global temperature are intimately linked (see figure 6).[14]

On one occasion around 135,000 years ago, atmospheric carbon dioxide rose to preindustrial levels, accompanied by a 5°C to 7°C rise in the average global temperature. Many climatologists claim there is even solid evidence to suggest that average global temperatures have fluctuated by as much as 10°C during the past million years and in some instances have achieved this climatic transformation in no more than a decade. On one such occasion the polar ice caps melted enough to raise sea levels about 6 meters (20 feet) above their present mark for a brief period.[15]

According to satellite measurements begun in 1978, the ice cap has been shrinking by 1.4% each decade, in response, it seems, to the 2.5°C rise in polar temperature recorded in Antarctica during the past fifty years. Should temperatures continue to rise at that rate (and there is no reason to suppose they will not), it seems doubtful that much of the present ice cap would survive the twenty-first century. If, on the other hand, positive feedback mechanisms (i.e., primary mechanisms that feed back into the system to trigger other contributive mechanisms) were to induce what is known as runaway warming and melt all polar ice—as has happened many times in the past—it would add some 25 million cubic kilometers (6 million cubic miles) of water to the world's oceans, raising the sea level by about 65 meters (200 feet), drowning not only the most densely populated regions on the planet but most of the world's richest agricultural land.[16]

Problems

Runaway Warming

However unlikely this scenario might seem, we should not dismiss it lightly. As the oceans continue to warm and expand and the earth's polar ice caps continue to melt, large areas of densely vegetated tropical lowlands are certain to become inundated. The methane (swamp gas) that this will generate represents a feedback factor of ominous potential.

Another methane threat lies waiting far to the north in the frozen wastes of the world's tundra regions. The melting of a single cubic meter of permafrost may yield up to three times that volume of methane gas through the decay of deep-frozen peat, which it generally contains. Even more dangerous is an icy mix of organic gases known as gas hydrates that are known to underpin some tundra regions to a depth of more than 400 meters (1,300 feet). First recognized in deep-sea sediments, these gas hydrates appear to form only in high-pressure, low-temperature environments where little or no oxygen exists. They appear in drill cores barely 50 meters (about 160 feet) from the surface. When thawed, a single cubic meter of hydrate may release more than 160 times that volume of methane gas.[17]

The vast reserves of methane trapped in gas hydrates represent the accumulated wastes of an ancient group of bacterial organisms known as archaebacteria, or archaea. Methane absorbs radiant heat at wavelengths not absorbed by other greenhouse gases and on average is some twenty times more effective as a heat retainer than carbon dioxide.[18] The methane now locked in the world's tundra regions therefore represents a carbon reserve of the most dangerous kind. If the permafrost begins to thaw, releasing methane initially from biological decay near the surface and then from the gas hydrates lower down, it will add a feedback factor that will bring runaway warming well within the realm of possibility. In northwestern Canada and Alaska this has already begun to occur. Average temperatures in the region have risen 1°C per decade over the past thirty years, causing the permafrost to melt in many places. This leaves the landscape pockmarked with holes called thermokarsts where underground ice masses have melted. Some 2,000 landslides have been attributed to the collapse of the permafrost.[19]

Overlaid on our current greenhouse emissions, a significant release of natural methane represents one of the most potent threats faced by our species. All of the temperature increases recorded in the ice cores are from five to fourteen times greater than would be expected from the increases in carbon dioxide that accompanied them, so it seems certain that some kind of positive feedback mechanism occurred, flipping the global climate from a glacial to a nonglacial state as though it were a binary system. According to scientists who analyzed the air-bubble record in the 2.5-

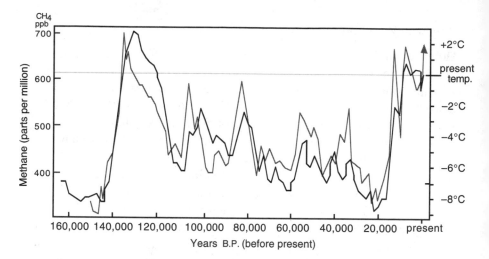

Figure 7. The temperature-methane link. The close match between drill-core methane levels (shown in gray) and temperature fluctuations (in black) during the past 150,000 years supports the proposition that methane is the trigger for runaway warming events such as occurred almost 140,000 years ago and about 15,000 years ago. Methane levels are shown in parts per billion (ppb), and the temperature in degrees above and below the present mean polar temperature. (Based on the Vostok ice-core data.)

kilometer (1.55-mile) ice core extracted from the Vostok drill site in Antarctica, "the fundamental link between methane and climate variations was unequivocal," indicating that "the natural methane cycle may well provide a positive feedback in any global warming" (see figure 7).[20]

As borne out by the ice-core data, global temperature and atmospheric methane levels appear to have been relatively stable for most of the past 10,000 years. That stability came to an end about two hundred years ago. Since then, according to the Vostok ice core, the amount of atmospheric methane has more than doubled and is now rising fifty times faster than at any time in the past 300,000 years.[21] This rise seems to be primarily attributable to the expansion of flood irrigation in tropical areas and to our increasing herds of flatulent cattle. In other words, the methane we are adding to the atmosphere is an inherent byproduct of human food production, and given our impending food crisis, is not therefore reducible to any significant degree.

Twelve Indicators of Global Warming

1. As displayed in polar drill cores, a close correlation has existed between atmospheric carbon content and air temperature for more than 200,000

years. Polar drill cores also show that carbon dioxide and methane levels rose sharply in the twentieth century.

2. The mean global temperature has increased by about 0.6°C in the past hundred years and is now rising faster than ever.[22] The last three decades have been the most climatically unstable, with the warmest years occurring in the 1990s.

3. The anomalous pattern of equatorial wind and water currents in the Pacific region, known as ENSO, has reappeared with increasing frequency in the last three decades. Similar atmospheric disturbances have been recognized in the Atlantic and Indian Oceans, and although less severe, they too appear to be linked to rising air temperatures.

4. Average air temperatures over Antarctica have risen by 2.5°C during the past fifty years, and the average water temperature of the Southern Ocean near Macquarie Island has risen by 1°C during the twentieth century.[23]

5. In response to polar warming, the James Ross and Wordie ice shelves that once attached to the northern and western coast of the Antarctic Peninsula have largely disintegrated. More recently, the huge Larsen ice shelf that flanks the peninsula's eastern side for 1,000 kilometers (620 miles) developed a network of giant fractures that signal its gradual disintegration, and in April 1998 a slab of the Larsen shelf 40 kilometers long and 5 kilometers wide (about 77 square miles) became completely detached.

6. Analysis of whaling records shows that Antarctica's vast apron of sea ice shrank by 25% in the twentieth century, shedding almost 5 million square kilometers (2 million square miles) in the last fifteen years of legal whaling (1972–1987).[24] According to satellite measurements the shrinkage rate is now 1.4% per decade.

7. Sea ice is disappearing even faster in the Arctic. Satellite measurements show that the melt rate increased from 2.5% to 4.3% per decade during the latter half of the twentieth century; vast stretches of the Arctic Ocean have warmed by 1°C or more in the past decade alone.[25]

8. Alpine glaciers on all continents have lost almost half their volume since 1850, and snow lines have retreated some 200 meters (about 650 feet). Similarly, the world's few equatorial glaciers will disappear entirely within a few decades if their present rate of recession (45 meters, or 150 feet, per year) continues.[26]

9. Sea levels are now more than 15 centimeters (6 inches) higher than they were at the beginning of the twentieth century due to thermal expansion and the addition of meltwater from alpine glaciers and polar ice caps. The current increase of 2 millimeters per year suggests that the process is accelerating.[27]

10. Coinciding with rising water temperatures in the eastern Pacific, declines of up to 80% have been recorded in its levels of zooplankton, posing a serious threat to the ocean's viability as a carbon sink. Plankton enable the world's oceans to absorb between 20% and 40% of the carbon dioxide that is added to the atmosphere each year by human activity. Warming causes thermal stratification of the water and prevents nutrient-rich abyssal currents from surfacing. Without this food supply, photo-plankton die, as do the zooplankton that eat them.[28]

11. Large areas of shallow-water coral reefs are bleaching and dying as sea temperatures climb past the upper limit of tolerance for the coral's colorful symbiotic algae. When the algae vacate the coral, the polyps die. More than 10% of the world's tropical reef systems are now affected in this way.[29]

12. Aided by unprecedented tropical drought, record temperatures, and an accumulation of tinder-dry forest litter, "more tropical forest burned in 1997 than at any other time in recorded history."[30] In 1998 the rate of burning accelerated.

Although such signs of biospheric distress might seem ill-defined and insignificant compared to the immediate threats implicit in the daily cut and thrust of national and international politics or the international stock market, the fact remains that those things only help to define how we live; it is the environment that ultimately decides whether we live. Since climate is a primary factor in determining the carrying capacity of human habitats, climate stability is essential for the maintenance of large populations. The global climate has been relatively stable for 10,000 years. Should that stability disappear, the 6 billion people that now inhabit this planet would no longer be able to feed themselves. Yet, as geologist Wallace S. Broecker of Columbia University once remarked, "Climate is an angry beast, and we are poking it with sticks."[31]

Land Degradation

Humanity feeds itself largely on the product of 36% of the world's available land surface. Of that, about 11% is devoted to growing crops and the rest is used as permanent pasture.[32] If you discount the tundra and the desert, the rest is either under valuable forest, is poor in natural nutrients, or has been hopelessly degraded by previous human use. Many scientists therefore argue that the continuing degradation of the world's limited productive land represents the most immediate threat that the swelling human population now faces.

Artesian water and cattle in central Australia. Large herds of cattle in central Australia have had a withering effect on the environment, tearing the vegetative cover to shreds and churning the thin topsoil to dust.

Land degradation falls into four general categories: soil exhaustion, erosion, salinization, and waterlogging. Thousands of hectares of good farming land are lost to agriculture each year for these reasons, and in many cases the losses are permanent. The first three contribute in varying degrees to a general process known as desertification, in which formerly productive land drifts gradually toward sterility. Extreme examples of desertification are readily apparent in many regions of Africa, Mongolia, China, southern Russia, and the Middle East, while the general symptoms of it are now clearly visible in the southern United States and Australia. Such permanent losses are serious wherever they occur, but in densely populated regions such as Africa and China they are catastrophic. Erosion and desertification now threaten 40% of Africa's nondesert land and 30% of Asia's productive area, while the global loss of cropland is currently running at 5 to 6 million hectares (19,000 to 23,000 square miles) a year, or 4% each decade. It is for these reasons that China's efforts to feed one-fifth of the world's population from one-fifteenth of the world's arable land has produced a rate of cropland destruction that is double the global average.

Desertification in central Australia. Hard-hooved grazing stock, supported by abundant reserves of artesian water, have set this region of central Australia on the path to total desertification. Up to 70% of Australia's rangeland has been affected in this fashion to some degree.

Similarly, the former Soviet Union, by means of "agricultural collectivization" and chronic mismanagement, has succeeded in destroying an astonishing 20% of its formerly productive grain lands in less than two decades.[33]

More than 70% of the world's rangelands have also been significantly degraded by overstocking and other inappropriate farming practices. Australia provides a textbook example of this. If it had not been for the discovery of the massive aquifers that underlie much of the continent, most of its arid zone (about 70% of the total area) would have remained unleased and safe from the huge herds of hard-hooved stock that have been allowed to trample its fragile ecologies for the past 150 years.

The legacy of all this earnest malpractice is that during the last two decades more of the developed world's agricultural land became unproductive through exhaustion, desertification, salinization, and waterlogging than was newly brought under cultivation. As a direct result, and because of our continued population growth, the world's per capita food production has begun a steady decline that shows no sign of easing.[34]

Exhausted Soils

The more intensively land is farmed, the sooner the soil becomes depleted of essential plant nutrients. Deficiencies may be masked initially by adding more fertilizer, but this has limitations due to ever diminishing returns. In the 1950s each ton of fertilizer added an average of 45 tons of grain to the harvest. By 1965 the return per ton of fertilizer had halved. By the 1980s the return had halved again. The introduction of the high-yield "wonder-crops" of the 1960s and 1970s only accelerated the problem. A plant that grows faster or produces more seeds requires more nourishment, and the nutrients for that growth must come from the soil. Meanwhile the addition of artificial fertilizer supplies only a few of the necessary trace elements and other micronutrients. Consequently, as these become exhausted the yield declines, regardless of the added fertilizer.

Those who clear tropical rain forest to plant crops face similar problems. In contrast to the fertility implied by the abundant vegetation that rain forest soil produces, such soils are usually quite poor. Ash left by burning the cleared vegetation ensures that the first two or three crops are good, but the yield declines sharply thereafter, prompting the farmer to clear more forest and begin the sequence again. This scenario, combined with rapacious commercial logging, is helping to destroy the last of the world's great tropical forests in Brazil, Borneo, and Thailand. Meanwhile the strategies that are used to prop up modern high-yield crops and overstocked pastures, such as irrigation and the liberal use of pesticides and fertilizers, only add to the ecological time bomb by helping to accelerate the rate of soil exhaustion and the loss of biodiversity on all continents

Erosion

In the world's arid regions almost all of the soil's meager nutrients are concentrated in the top centimeter. Any human activity that exposes this layer to increased erosion by wind and water represents an open invitation to disaster, since the rate of soil erosion escalates alarmingly the moment the crust is disturbed. Australian research into arid-zone erosion shows that even on the best managed agricultural land the rate of soil erosion may be many times greater than the rate at which new soil is formed. When the vegetative cover or algae-bonded crust is broken by grazing stock or farm machinery, the crucial upper centimeter of soil, which may have taken up to a thousand years to develop, can disappear in less than a decade. The subsequent decline in native vegetation then results in lower rainfall, increased erosion, and the general drift toward sterility that characterizes the desertification process.

These factors now contribute to an annual loss of about 6 million hectares (23,000 square miles) of productive cropland, or 4% of the global total, every decade. Because the effect is cumulative, it represents a massive reduction in our future capacity to feed ourselves. According to a 1990 United Nations survey, about 1% of the world's topsoil is lost to production every year through wind and water erosion. This loss, somewhere around 25 billion metric tons, is roughly the same volume of topsoil that underpins Australia's extensive wheat lands.[35]

The countries showing the most erosion damage are Africa, China, India, the United States, and Australia. It is worth noting, however, that Africa's population is around 700 million, China's 1.2 billion, India's almost 1 billion, and the United States' 250 million. With a mere 18 million inhabitants, Australia seems out of place, and its inclusion in this list requires some explanation.

Not only is most of the Australian continent extremely arid and its soils extraordinarily poor, but Australian farmers have been particularly proficient at extracting whatever resources the beleaguered soil has to offer. Moreover, the introduction of a wide variety of very damaging exotic species such as cattle, sheep, goats, pigs, buffalo, camels, donkeys, rabbits, cats, and foxes has also had a devastating effect. In fact, the deadly partnership of agricultural malpractice and introduced species has saddled Australia with some of the highest rates of erosion in the world.

In the arid, overgrazed Western Division of New South Wales, for example, annual erosion rates (by wind and water) of up to 200 metric tons per hectare (81 tons per acre) have been recorded. A single dust storm, such as the one that ravaged the South-West Land Division of Western Australia for seven hours in May 1995, can strip cropland of up to 3 millimeters (about one-eighth of an inch) of topsoil at a time, carrying off in one day millions of tons of soil that may have taken thousands of years to form. The dust cloud generated by a storm that raged across drought-stricken southeastern Australia in May 1994 measured 800 kilometers (nearly 500 miles) long, 200 kilometers (120 miles) wide, and 200 meters (about 220 yards) deep. Tracked by satellite, it was calculated to contain some 20 million metric tons of topsoil. Most of that precious load was finally dumped into the Tasman Sea.[36]

Paradoxically, it was an abundance of water, not on the surface but underground, that led to Australia's disastrous record of land management. In dramatic contrast to the surface aridity, vast reserves of water are locked in the giant artesian basins that underpin most of Australia. One of these, the Great Artesian Basin, underlies more than one-fifth of the continent and with an area of about 1.7 million square kilometers (about 650,000 square miles) appears to be the largest in the world.[37] This aquifer has been so drained of its water by the 4,000 bores that were drilled into it in the early part of the twentieth century that 1,000 of them have now ceased to flow. The water that has gushed uselessly for decades into the desert and semidesert that surrounds most of these bores took between 1 and 2 million years to accumulate. In anthropocentric terms it is irreplaceable.

Tapped by some 25,000 artesian and subartesian bores, Australia's gigantic subterranean reservoirs made it possible to run large herds of thirsty, hard-hooved grazing stock over most of the continent, even in the driest of years. During such times this inflated army of wholly unsuited herbivores strips the country bare of its protective grasses, breaks up the algae-bonded crust with their hooves, and invites the wind to remove what little nutrient remains in the impoverished topsoil.

The problem is now widely recognized, and yet about 6,500 square kilometers (more than 2,500 square miles) of virgin land are still cleared every year. This is more than half the rate at which the tropical rain forests of Brazilian Amazonia are being cut down. Worse still, the burning and decay of the vegetation uprooted during land clearing adds significantly to Australia's disproportionate greenhouse gas emissions. In 1990, for example, land clearing added 155 million metric tons of carbon dioxide to the atmosphere and accounted for 27.3% of Australia's net greenhouse gas emissions. It is one of the main reasons Australia's 18 million people are also the world's most prolific greenhouse gas producers on a per capita basis and contribute between 1% and 2% of the total global emissions.[38]

Bearing in mind that the average Australian—like the average American or western European—consumes more commercially produced energy and resources in a year than do 120 average inhabitants of a country such as Bangladesh, it can therefore be argued that the Australian continent would fare no worse were it occupied by 2 billion Bangladeshis. Underpopulated it is not.

Subjected to ever increasing human assaults of this kind, natural ecologies have been gradually collapsing around the globe for thousands of years. When a region became untenable in the past, however, its inhabitants usually had the option of moving elsewhere. With land degradation now occurring on a global scale, the problem assumes a very different proportion.

The Salinity Time Bomb

Soluble salts are a major component of the substrata in many regions of the world and may be brought to the surface when water tables rise. The long taproots of trees act as pumps and usually prevent the salt from rising by keeping the water table down. But should those natural pumps be

Lost River, Great Sandy Desert, northwest Australia. As one of the world's oldest and flattest continents, vast areas of Australia have been inundated by the sea at various times. This spring-fed, salt-encrusted river in northwestern Australia betrays the subterranean salinity that now causes so much agricultural hardship in other parts of Australia.

removed to make way for stock or crops, the land may eventually lose its arability due to the accumulation of salt at the surface.

This process, known as dryland salinization, has been going on for thousands of years all around the world, most notably in the Middle East, India, and China, and has already rendered large tracts of formerly productive land infertile. Australia, the world's driest inhabited continent, is one of the newest victims and one of the worst affected. About 2 million hectares (more than 7,700 square miles) of farmland are now salt damaged, and another 1 million hectares (3,860 square miles) are under imminent threat. The consequent loss in food production is currently estimated to be around $300 million a year and mounting.[39]

Another, more direct route to that final stage of salinization is irrigation. All water that has been in contact with the earth inevitably carries a small percentage of salts in solution. Irrigation by spray, drip-feed, or flooding invariably leaves some of this mineralized moisture on the soil surface. Evaporation of the water means the continual addition of salt residue to

Cotton irrigation in New South Wales, Australia. Massive water consumption by cotton growers based on the headwaters of Australia's Darling River have so consistently depleted the flow of this, the continent's only major river system, that its downstream ecologies have virtually collapsed. Loaded as it is with fertilizer, pesticides, herbicides, and defoliants, water that has been used for irrigating cotton fields is generally unfit to be returned to river systems.

the soil, eventually making the land unsuitable for growing any traditional food crops. To make matters worse, most of the world's modern irrigation projects have been located in areas so arid that less than 50% of the impounded water is ultimately absorbed by the crop: the rest is lost to seepage and evaporation. Not only does this constitute a salinity time bomb, it also represents massive waste of an arid region's most precious resource. Between 1977 and 1997 the world demand for freshwater tripled as a result of rising population, increasing irrigation, and the special requirements of high-yield "green revolution" crops.[40]

Around the world some 400,000 square kilometers (about 155,000 square miles) of formerly productive agricultural land is currently underused or has been abandoned due to high salt concentrations, and the rate of loss is rising. Despite these losses, the global percentage of land watered by irrigation has more than doubled since 1950, and one-third of the world's harvest now depends on irrigation to some extent.[41]

The world's most catastrophic example of irrigation related salinization is the Aral Sea basin in the Republic of Uzbekistan. Once the fourth largest body of inland water in the world, the Aral Sea has shrunk to less than one-third of its original volume and half its original area in just forty years, its shorelines retreating up to 120 kilometers (72 miles) in places. This vast inland sea once yielded up to 50,000 metric tons of fish a year, providing the region's inhabitants with their primary source of protein. The sea's last fish died in 1983 and most of its former fishing ports now lie out of sight of the shoreline. If the planned recovery program fails, as seems likely, Russian scientists expect that the rest of this dead sea will vanish altogether by the year 2010, leaving only a vast, salt-ridden wasteland.[42]

The former Soviet Union used to produce about 68% of its cotton through intensive irrigation in this region. The cotton farms used so much water that for twenty to thirty years almost none reached the Aral Sea from the surrounding mountain catchments. The residues of pesticides together with remnants of the Soviet cotton defoliant Butifos, which was used liberally on the cotton crops until the late 1980s, have contaminated all the natural water resources in the region, creating an irreversible tide of disease and death for the inhabitants and their livestock. The use of Butifos is now banned, but the region's high incidence of kidney and thyroid diseases and esophageal, stomach, and liver cancers continues to climb, while viral hepatitis and tuberculosis affect 50% more people than in 1985. Of the 700,000 women in the Karakalpakstan region, some 97% now suffer from severe anemia and are presently giving rise to an entire generation of anemic children who are, in turn, doubly prone to other infections.[43]

One might think that the close link between irrigation and salinization had only recently been discovered, yet salinization appears to have played

Waterlogged pasture in Murray Valley, New South Wales, Australia. Extensive tree clearing has contributed to a general rise in agricultural water tables throughout eastern Australia. In parts of the Murray River basin, water tables have been rising more than 30 centimeters (1 foot) a year, bringing both water and salt permanently to the surface in many places.

a leading role in the decline of the world's oldest and most successful civilizations, the Sumerian culture of Mesopotamia being the most notable example. Having given birth to written language, mathematics, written law, and irrigation, this civilization appears to have finally surrendered to the insidious creep of salinization about 4,400 years ago. Despite improved farm techniques and the introduction of salt-tolerant crops such as barley, an annual decline of less than 0.1% over a period of seven hundred years eventually rendered all irrigated cropland infertile, causing the civilization to crumble and the people to disperse.[44]

Waterlogging

Where the water table is naturally high, the removal of trees may result not merely in a chronic case of rising dampness each winter but the permanent waterlogging of the soil and the total loss of agricultural productivity. Overzealous land clearing has achieved this dismal result on many continents, even arid ones like Australia. Recovery of the affected land is frequently impractical, due not only to the high cost of drainage and reforestation but also to increased salinization.

Waterlogging is also caused by the lack of adequate drainage systems in irrigated areas and the overworking of clay-based soils. Both of these factors tend to build an impervious clay layer immediately below plow level, leading at best to a significant reduction in crop yield and at worst to permanent loss of the land to any form of agriculture.

Agricultural Limits

The primary mechanism of population control throughout the animal world is the availability of food. Our species is no exception. According to the FAO, the world currently produces just enough food to feed each of the world's 6 billion people an adequate low-protein diet of 2,500 calories a day. Although the current rate of population growth is still about 80 million a year, the world's food production began to falter in 1984 and has barely kept pace with demand since then. To make matters worse, most of the world's arable land is already in use and its yield is shrinking annually due to various forms of degradation.[45]

The key figures are those for grain production, which peaked globally on a per capita basis in the mid-1980s and has since declined by an average of 1% a year (see figure 8).[46] The 1957 fall was probably the first in history, but it has since declined several times in response to the worldwide environmental stress caused by repeated ENSO disturbances. In gross tonnage, too, global grain production has fallen on several occasions (see figure 9).

In recent years the world's only reliable food exporters have been the United States, Canada, the European Common Market, Australia, New Zealand, Argentina, and Thailand. By far the most important of these regions is North America, which currently supplies about three-quarters of all the world's food shipments. A prolonged drought there is now of worldwide concern. When the United States' catastrophic 1988 ENSO-related drought began, global grain reserves would have fed the world for about 104 days. Just seven years later, at the end of the 1991–1995 ENSO, the most prolonged global climate disturbance on record, global grain reserves had fallen to just 231 million metric tons, enough to feed the world for only 48 days according to Lester Brown, president of the World Watch Institute.[47] With agricultural production close to its limits and the per capita grain harvest declining, humanity finally faces the constraint that ultimately keeps all plague species in check: shortage of food.

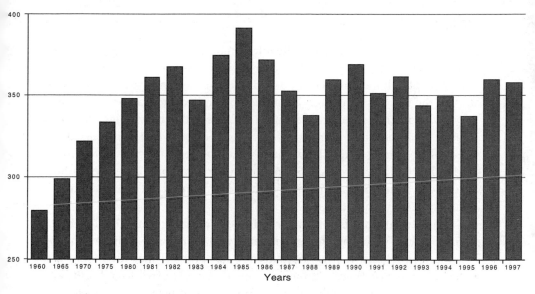

Figure 8. World grain harvest (per capita). (*Source:* United Nations Food and Agriculture Organization.)

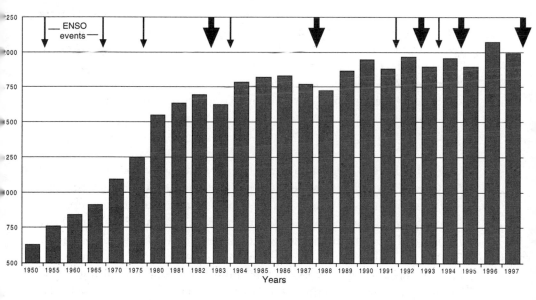

Figure 9. World grain harvest (gross tonnage). (*Source:* United Nations Food and Agriculture Organization.)

The Green Revolution

Had the much vaunted Green Revolution of the 1960s and 1970s not come to our rescue, the human population explosion would likely have come to an abrupt halt more than three decades ago. Between 1950 and 1984, however, the world's annual grain harvest multiplied by a factor of 2.6, while population grew by a factor of only 1.9. Little of this harvest increase came from the addition of more cropland. Most of it was due to increased irrigation and the introduction of high-yield species, nitrogenous fertilizers, and synthetic pesticides.[48] Science won for humanity an eleventh-hour reprieve from the food crisis that had been predicted by Paul Ehrlich, Julian Huxley, and others.

Unfortunately, this reprieve reinforced the dangerous illusion that a combination of technology and ingenuity could provide solutions to most human problems. It was not generally understood that high-yield also meant high cost. Since high-yield crops inevitably extract more nutrients from the soil, they require heavy applications of fertilizer and pesticide, which means more frequent soil disturbance, more erosion, compaction, acidification, and a growing imbalance in the remaining micronutrients and trace elements. These stresses on the soil further reduce fertility, prompting the application of additional fertilizer just to maintain the same return. Hence more erosion, compaction, and so on. To make matters worse, the introduction of fast-growing species enabled farmers to plant two or even three crops a year, thereby multiplying the rate of damage. In other words, the real price of artificially induced harvest abundance is degraded land, and the larger the harvest, the greater the environmental debt, a debt that will likely be paid in most cases by our children.

A Nitrogen Fixation

One of the keys to the leap in agricultural productivity that accompanied the Green Revolution was the mass production of nitrogenous fertilizers. Artificial synthesis of ammonia, the process that underpins the fertilizer industry, was invented by two German scientists, Carl Bosch and Fritz Haber, at the beginning of the twentieth century. In the words of Vaclav Smil of the University of Manitoba, Canada, "At least two billion people are alive because the proteins in their bodies are built with nitrogen that came—via plant and animal foods—from a factory using [Bosch and Haber's] process. . . . In just one lifetime, humanity has indeed developed a profound chemical dependence."[49]

There are drawbacks to this dependency. Overfertilization with nitrogen-based compounds acidifies the soil and water and allows nitrogen to return to the atmosphere as nitrous oxide. There it contributes both to the destruction of the ozone layer and to greenhouse warming. Fertilizer residue that washes into the world's rivers, lakes, and oceans constitutes a major influx of a previously scarce nutrient, which leads in turn to massive algal and cyanobacterial blooms that deoxygenate the water, in some cases triggering a general collapse of the regional biota. Modern human food is anything but cheap.

The Energy Debt

Boosting agricultural productivity by technological means merely constructs an accelerating treadmill for the farmers, with no soft landings for those who want to get off. High-yield crops, fertilizer, herbicides, pesticides, fungicides, and sophisticated agricultural technology create nothing. They merely borrow from the global energy bank, and we are forced to accept these loans in order to meet ever-increasing consumer demands. These are short-term, high-interest energy loans, which our growing human family can now barely service and cannot hope to repay. They represent an energy mortgage that will eventually cost our grandchildren the entire farm.

This massive debt is compounded by the tendency of intensive agriculture and high-yield crops to actively encourage assaults by pests and diseases. Of the food produced around the world each year, somewhere between 25% and 40% fails to reach consumers due to spoilage, and up to three-quarters of this loss is due to damage by pests and diseases.[50] The understandable response of the beleaguered farmer has been to apply more chemical pesticides. However, most of these chemicals attack not only the problem species but their natural predators as well, and since predators are generally longer lived and much less fertile than their prey, the collateral destruction of predators during one season invariably means even greater pest problems for the farmer next season. And so the dismal cycle perpetuates itself.

The Perils of Biotechnology

There appear to be only two avenues of escape from the fertilizer-pesticide treadmill. However, both involve sophisticated biotechnology and are still in the development stage. The aim is to build pest protection into the fabric of high-yield plants, either by means of genetic engineering or by employing hormone sprays that are absorbed into the tissue of the plant or its seeds. Although all the long-term consequences of these strategies are not well understood, at least they leave the existing population of natural predators unharmed. Since insects and other pests grow more immune to chemical pesticides with every passing year, this defense offers the most realistic prospect of feeding the continually growing human population.

If such high-tech strategies live up to expectations, they will postpone the global food crisis yet again, perhaps for another couple of decades if we are very lucky. However, the business of tweaking the molecular levers of life is so complex and so riddled with pitfalls that the long-term consequences are virtually unforeseeable. With the stakes so high and our knowledge so limited, the risks are enormous. In fact, the first hint of danger has already appeared. Recent laboratory experiments in Canada and the United States have shown that genetically emasculated viruses used to immunize crops have sometimes reacquired their missing genes from transgenic host plants that happen to carry them.[51] This means that there is also a significant risk that wild viruses will hijack viral material from genetically immunized crops and thereby create mutant strains that are already immune to the latest herbicides. This kind of genetic free trade certainly occurs between viruses in the wild. DNA analysis has shown that two cassava plant viruses recently pooled their genetic resources in Uganda to produce a more virulent strain that is now decimating entire crops of this staple food.[52] In other words, transgenically altered crops—those that carry an emasculated form of a pathogenic species—tend to become the incubators of new, more resistant diseases.

The FAO reported in 1996 that the survival of the entire European barley crop now depends on just one gene and one fungicide, while a virulent new strain of the fungus known as potato blight has recently developed a countermeasure to the genetic protection that has been bred into potatoes to resist the fungus. Even more ominous, the new species has also become resistant to all known fungicides and has demonstrated an ability to mutate rapidly to combat new resistance genes and new fungicides.[53]

The development of such genetically versatile pests and diseases gives clear warning of the dangers of monoculture. About a million people starved to death in Ireland when potato blight wiped out Ireland's potato harvest in 1845, 1846, and 1848. The massive death toll was a direct consequence of Ireland's dependence on this one crop, yet increasing monocultural crops now seems unavoidable. Following the granting of a seed patent to the U.S. Department of Agriculture and a Mississippi seed firm in March 1998, all U.S. seed companies are now free to develop and market agricultural seeds of all kinds that will grow normally but yield only sterile seed in the harvest, meaning that farmers will have to buy the transgenic seed for each new sowing.[54] Half the world's farmers are too poor to buy commercial seed and rely on replanting with their own harvested seed. This currently supplies between 15% and 20% of the world's food and preserves whatever genetic diversity resides in their crop. Presterilized seed, on the other hand, not only minimizes genetic variation but also represents a financial bonanza to commercial seed companies everywhere. Provided governments fail to ban such seed, it should spread quickly around the world, especially in the wake of ENSO droughts, thereby binding humanity to an even narrower and more vulnerable food base and sending millions of subsistence farmers deeper into debt—or starvation.

This quagmire serves to illustrate yet again that technological solutions are illusory, and even the most sophisticated biotechnology and adroit molecular manipulation will inevitably conceal costs that ultimately negate the benefits delivered. Once again, "high yield" simply means "high cost." Evolution has always depended on it. As if in pessimistic confirmation, governments around the world have begun to scale down their funding for agricultural research, and with economic rationalism and religious fundamentalism gaining ground everywhere, this trend seems certain to continue.

Marine Harvest Decline

As on land, the harvest from the sea has reached an ominous plateau (see figure 10). The global catch peaked on a per capita basis in 1970 and has stagnated ever since (see figure 11). According to many authorities, the industry now faces imminent decline due to a combination of overfishing, coastal habitat destruction, and water pollution. The FAO reported in 1996 that 25% of the world's commercially useful fish stocks "are now overexploited or have already crashed, while another 44% have reached their limit."[55] Only in the Indian Ocean is the catch still increasing. The FAO's 1995 report warned of a worldwide crisis that threatens everything from plankton to porpoises, with the odds especially stacked against the top predators, including such vital food species as tuna, cod, and herring. With more than a million fishing boats now operating, and new ones still being built, the situation is unlikely to improve.

Rapacious harvesting methods, satellite positioning, and sophisticated fish-detection systems employed by fleets of big, fast refrigerated trawlers—floating factories that can catch and process a ton of fish every hour—have brought the strategies of open-cut mining to the world's oceans with catastrophic efficiency. The world's richest cod fisheries, on the Grand Banks off Newfoundland, totally collapsed in 1992 due to overfishing and the practice of dragnetting.[56] Similarly, the vast shoals of salmon, herring, and tuna that once characterized the northern Pacific and

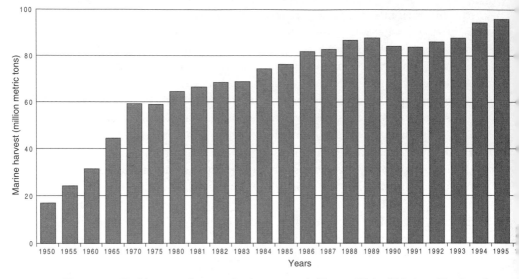

Figure 10. World marine harvest (gross tonnage). (*Source:* United Nations Food and Agriculture Organization.)

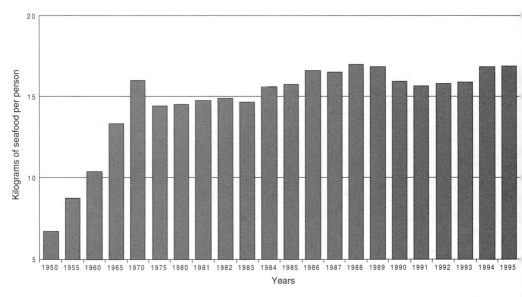

Figure 11. World marine harvest (per capita). (*Source:* United Nations Food and Agriculture Organization.)

Atlantic Oceans have all but disappeared, and Japanese and Russian fishing boats now regularly head for the southern hemisphere at the end of the northern season in order to fill their quotas.

Australian fisheries are showing similar signs of decline due to habitat destruction and overharvesting of their poorly understood resources. With its meager rainfall and few rivers, Australia presents a continental shelf that is, by marine standards, rather like the continent's arid heart, a fragile semidesert. Overfishing consequently tends to produce disproportionately savage declines in its fish stocks, and Australia's annual catch has steadily decreased since its 1991 record of 157,100 metric tons.[57]

Aquaculture

The great hopes that were once raised for fish farming have also evaporated. Although the practice of stocking lakes, reservoirs, and marine pens with marketable fish and shellfish has great potential for growth around the world, this too entails a massive hidden fee. Since maintaining and feeding large stocks of marketable fish is an expensive business, and because the product is predictable and generally of high quality, the industry is understandably geared toward supplying the high end of the market. In other words, the poor miss out. To feed captive fish—generally a predatory species such as salmon, tuna, or trout, or even scavenger species such as prawns and crabs—fish farmers must buy large quantities of fish meal made from wild pelagic species such as anchovies and sardines. Scavengers are not fussy eaters, but even so it takes more than two tons of fish meal to produce one ton of shrimp.[58] If you switch to farming fussy eaters such as salmon, however, your fish-meal order will triple. In other words, aquaculture is a process by which large quantities of low-quality wild fish are turned into a small quantity of high-quality, high-priced farmed fish. By interposing a so-called production stage in the food chain, we do not produce more food, but less. In reality, the fish farmer is just another fish consumer, and a big one at that.

Even more insidious, the construction of the large coastal pens required to hold the stock frequently entails the destruction of mangroves and other nursery habitats that are vital to the preservation of many wild species, ensuring that there will be even fewer wild fish to catch in the future. And finally, just like irrigation projects, fish farms have a limited life. In just five to ten years the accumulating fallout of unused food, fecal matter, and dead bodies eventually overwhelms the natural bottom ecosystem, causing eutrophication and forcing the fish farm to close and move to a new site. This limited-life factor is one of the major reasons that half the world's mangroves have now been destroyed. According to a 1996 FAO report, in just six years shrimp ponds consumed more than 17% of Thailand's mangroves, while in India the losses approach 40%.[59]

Once again, an "environmentally friendly" industry turns out to have hidden costs that make it more like a creeping disease than the blessing it is usually touted to be. In the sea, as on the land, those free lunches we continue to seek are illusory. Every activity exacts a commensurate environmental fee, and in our present state of global plague the golden age of natural abundance now lies far behind us.

Nutrition

According to a 1993 United Nations survey, global food production was sufficient to feed each human being on the planet about 2,500 calories a day.[60] That figure was some 200 calories above the UN's recommended minimum intake, but it is worth noting that the calculation was based on an essentially vegetarian diet. Had it been based on the protein-rich diet common in Australia, the United States, and western Europe—in other words one in which a quarter of the calories come from animal products—there would be enough food to feed no more than half the present global population.[61] The average Australian diet, for example, incorporates about 30% of calories from animal sources and is available only because the world's food is unevenly distributed. Almost a billion people have too little to eat. Underpinned by the Green Revolution and exemplified by grain production, per capita food production rose spectacularly between 1960 and the mid-1980s, but there has been a general decline since then. Hunger and malnutrition are more widespread than ever.

Meanwhile our taste for meat, fish, eggs, and dairy products exacts a high environmental toll. The conversion of solar energy first to edible biomass and then to animal protein is a slow, energy-expensive process. At each step in the food chain more than 90% of the original energy is consumed. For example, of the total incoming solar energy, less than 0.5% will be converted into edible biomass suitable, say, for a grazing steer, and only about 0.8% of the energy captured in the fodder will be installed in the muscles of the steer.[62] These figures are of course approximations, but the general principle applies throughout the world's food chains with roughly similar conversion rates (about 10%) at each step in the chain. Nevertheless we currently feed between a quarter and a half of our annual grain harvest to livestock, even though it would be far more energy-efficient to eat the vegetable biomass ourselves. To make matters worse, we then compound that extravagance by spending vast quantities of additional energy and resources in processing, packaging, transporting, advertising, and retailing the relatively small amount of animal-sourced calories that we do harvest.

In a dietary survey of more than 2,000 Britons, researchers from the Centre for Energy and Environment at the University of Exeter found that it cost, on average, around 18,000 megajoules of energy a year to get each consumer's food from the farm to the table. This was almost six times the energy contained in the food itself. Meanwhile the marketing process contributed some 15 million metric tons of carbon to the atmosphere in burned fossil fuels.[63]

Such extravagance is made possible largely because more than 1.1 billion people live in absolute poverty, and about 800 million of them are "food insecure"—which is economic double-talk for half starved.[64] The criterion for "absolute poverty" is that used by the World Bank and defined by its president, Robert McNamara, as "a condition of life so limited by malnutrition, illiteracy, disease, squalid surroundings, high infant mortality, and low life expectancy as to be beneath any reasonable definition of human decency."[65] Most of the world's poor live in developing countries, and during the next three decades 95% of the world's population growth will occur in those nations. The International Rice Research Institute in the Philippines calculates that unless the population cycle is broken in these regions, the global rice harvest will have to increase by 70% in that time just to maintain present consumption.[66] Yet, where population growth is greatest and malnutrition most prevalent, there is little room for agricultural expansion.

Even if we assume a declining birthrate due to social factors and a rising death rate due to starvation, disease, and genocide, the global population will grow by at least 1.2 billion by the year 2050.[67] Meanwhile the world's food producers have no more aces to play. The per capita marine harvest has stagnated since 1970, and fish stocks are now declining globally. We can no longer increase crop yields by using more fertilizer, nor can we add significantly to the harvest by cultivating new land, since most of the potentially arable land is of marginal quality at best, and clearing it usually means losing valuable forests. Moreover, the world continues to lose good farmland to degradation and urbanization much faster than new land can be brought into production.

We are left with few viable alternatives. We can increase the total area under irrigation and intensive cultivation, although as we have seen, this is a short-term fix that only puts us deeper into environmental debt. We can continue to introduce genetically "immunized" high-yield species that will increase the harvest by a small but vital annual percentage (and leave the environmental bill to our children). Or we might smooth out the gross disparity in wealth and food distribution, and thereby share what we have a good deal more equitably. And pigs might fly.

Global food production is approaching its limit, while the juggernaut of population growth rolls on.

Freshwater

The final arbiter of our planet's carrying capacity is the availability of clean freshwater. Humans currently siphon off 54% of all the world's readily available freshwater and pour more than two-thirds of that into agriculture. And since one-third of all our food is now grown on one-sixth of the available cropland—the area currently irrigated—water shortage is one of the planet's most pressing problems. According to a 1996 global water usage study, humanity's requirements, even assuming modest rates of population growth, will rise to about 70% of the earth's accessible freshwater by the year 2025.[68]

A major cause of this water shortage can be traced to our taste buds. Most of our favorite foods are astonishingly water-expensive. Recent studies in Australia and the United States show that producing a single kilogram (2 pounds 3 ounces) of beef takes anywhere from 50 to 150 tons of water, meaning that on average the meat in one large (quarter-pound) hamburger represents the consumption of about 11 cubic meters (approximately 2,900 U.S. gallons) of water. It is not that cattle drink a great deal, but rather that most of that water goes into producing their feed. And because cattle currently constitute the largest biomass of any mammal, they represent one of the planet's principal water consumers.[69]

Heading a research team with Australia's Commonwealth, Scientific and Industrial Research Organisation (CSIRO), Wayne Meyer of Charles Sturt University in New South Wales has similarly investigated the water consumption of some of the main crop species. Rice and soybeans proved to be the thirstiest of them, each guzzling between 1,900 and 2,000 liters (roughly 500 to 530 U.S. gallons) of water for every kilogram of edible seed that they produced. To put this in a nutritional perspective, 1 kilogram (2 pounds 3 ounces) of white rice delivers about 10 megajoules of energy to its human consumer. This is approximately the minimum required by an adult male to maintain bodily functions for one day and to fuel a reasonable level of physical activity. So even when humans are reduced to eating nothing but white rice, mere subsistence costs about 2 metric tons (almost 530 U.S. gallons) of water a day per adult.

Darling River, New South Wales, Australia. During the four-year drought of the 1990s, the lower reaches of Australia's Darling River, already compromised by massive irrigation schemes, thirsty stock, and growing urban needs, were occasionally reduced to a thousand-kilometer (six-hundred-mile) chain of toxic puddles.

Pollution

Most modern humans leave a trail of waste materials that is prodigious by earthly standards. Some of those wastes do little harm to the biosphere or to other life; some seem to be lethal to most species, ourselves included. An inevitable consequence of extravagant consumption, the amount of waste generated by humans is not unique in the spectrum of existence. What is unusual, though, is the diversity and dispersibility of our wastes. Riding conveyor belts of high-altitude air currents, many kinds of volatile pesticides and industrial pollutants are continuously redistributed from tropical sources to colder regions, where they accumulate in hitherto pristine landscapes. These chemicals, many of them highly toxic, are often vaporized directly from the soil and are driven poleward by the constant rise of warm tropical air in what amounts to a highly efficient global distillation process. Some of the world's most toxic pesticides, including DDT, chlordane, and toxaphene, head the list. Computer modeling suggests that the more volatile of them may be ferried from the tropics to polar regions in as little as five days.[70] Condensing out in colder latitudes, the chemicals then wash into river systems, where they become incorporated into the regional biota and accumulate in the body fat of the top predators, including humans.

The major contaminant in polar regions is toxaphene, a complex and volatile organochlorine that is widely used as a pesticide in tropical Asia and Latin America but is little used elsewhere. The research that uncovered this curious phenomenon was originally triggered by the discovery of extraordinarily high concentrations of toxaphene in the milk of Inuit mothers in northern Canada, even though use of the chemical had been banned in Canada for more than a decade.[71]

Ozone Destruction

In a similar pattern of poleward redistribution, several anthropogenic waste gases are now known to have attacked earth's thin but valuable shell of high-altitude ozone molecules. Ozone is an unusual form of oxygen, having three atoms to each molecule (O_3) instead of the usual two. This structure makes it inherently unstable. Although life-threatening to many organisms in other contexts, the ozone that has accumulated in the upper atmosphere during the past 2 billion years protects genetic material and its protein products from the sun's high-energy ultraviolet radiation. It was very likely the accumulation of this thin protective shell of ozone around the planet some 600 million years ago that finally allowed complex life forms to emerge from the sea and colonize the land. Although the usual sort of oxygen is also present in the stratosphere, it is comparatively stable and much less effective as a shield against ultraviolet radiation.

The threat to the ozone layer first appeared early in the 1980s when evidence began to emerge that chlorine molecules from some of our waste gases, particularly chlorofluorocarbons (CFCs), were attacking stratospheric ozone. Once CFC molecules rose more than 50 kilometers (30 miles) into the atmosphere, the high-energy wave bands of ultraviolet light (UV-B and UV-C) were able to break the chemical bonds that locked chlorine into the CFC group. Scientists investigating the phenomenon were astonished to discover that each molecule of chlorine monoxide released in this process was then capable of acting as a self-regenerating catalyst in the destruction of many thousands of ozone molecules, robbing an oxygen atom from each in turn.[72]

Recent studies have shown that other gases, such as methyl bromide, a chemical commonly used as a pesticide and fumigant, and nitrogen oxides discharged by high-flying jet aircraft, also significantly weaken the ozone layer, enabling UV-B to penetrate the atmosphere more easily. Fortunately, high-energy UV-C is not yet able to penetrate all the way to earth's crust. Under laboratory conditions, UV-C is known to attack many proteins as well as the genetic polymers RNA and DNA that govern all earthly life. The results of UV-C rays penetrating the stratosphere could well be devastating indeed.

The research that uncovered the process by which ozone was being destroyed was originally triggered by the appearance of a small hole in the ozone layer over the South Pole in the spring of 1982. By 1996 the hole was reappearing annually and had expanded to cover some 20 to 24 million square kilometers (about 15 million square miles), or roughly the area occupied by the Antarctic continent.[73] The rapid growth of this hole, which reappears each spring and lasts for several months, is reflected in the increasing incidence of skin cancer and cataracts and other eye damage in Australia and New Zealand. Ultraviolet-B also appears to depress the human immune system to the point that infectious diseases are more readily able to establish themselves. Laboratory evidence also shows that increased levels of ultraviolet radiation will attack marine photoplankton, thereby threatening not only the massive marine biota that depends on them as a food source but also the crucial marine loop of the carbon cycle on which almost all life ultimately depends.

Should it continue to increase, ultraviolet penetration will also begin to affect global food production by reducing the yield from some of the world's major crop species, such as rice, wheat, and soybeans. Laboratory experiments show that when such plants are subjected to a 25% increase in UV-B radiation their yields tend to fall by a similar percentage. Worst of all, the damage is not confined to polar regions, as was first thought. At midlatitudes, between 40° and 60° north and south, ozone depletion has been occurring at a rate of 4% to 5% per decade since 1979 and shows little sign of easing despite a recent reduction in CFC emissions in most Western nations. In fact, major losses are still being recorded, even at low latitudes. According to 1994–95 measurements the ozone levels over the Hawaiian Islands (about 21° north) had fallen by an astonishing 13%.[74] Meanwhile the global rate of loss continues at about 3% per decade.

Acid Rain

Sulfur dioxide, the most damaging component of acid rain, comes from a wide variety of sources both natural and artificial. About half of the atmosphere's sulfur originates during the burning of fossil fuels. The other half comes from natural sources such as sea spray, marine plankton, rotting vegetation, and erupting volcanoes. These combined emissions create an

Smog in Sydney, Australia. Suffocating beneath thick blankets of photochemical smog, large modern cities not only have become major sources of atmospheric carbon and acid rain, but they also provide breeding grounds for a growing variety of respiratory diseases.

environmental problem because of sulfur's tendency to react with hydroxyl radicals in the atmosphere to form sulfur trioxide gas. Being soluble, this gas combines with water vapor to form a weak sulfuric acid and eventually falls to earth as acid rain.

The ratio between natural and anthropogenic sulfur varies considerably from place to place. In some parts of Europe, human activity accounts for as much as 85% of the sulfuric acid in the rainwater. Acid rain, in combination with nitric acid formed from nitrogen oxides (byproducts of fossil fuels), acidifies soils and leaches magnesium from them, thereby blocking plants' ability to photosynthesize their food. According to a 1997 United Nations report on natural resources, some 60% of Europe's commercial forests have now been affected in this fashion to varying degrees, and in some countries, notably Poland, entire forests have died because of it. But the damage does not stop there. The runoff from acid rain also poisons fish and other aquatic life and is now a major source of water pollution in Europe and the United States. Some European river systems are now virtually dead because of the accumulating acidity.

Not all of Europe's acid rain comes from local car exhausts or Europe's sulfur-rich coals. A digital imaging system, based on reflected laser light and used on board the space shuttle *Discovery* in 1994, revealed a vast plume of sulfate aerosols that stretched from the industrialized eastern seaboard of the United States right across the Atlantic to Europe, where it blended with local emissions. There were even parts of Europe where acceptable levels of acidity were exceeded with no European input at all.[75]

Although the United States remains the main polluter of the world's atmosphere, it now has a close rival on the far side of the Pacific Ocean. China's rush to industrialize has produced a huge new crop of coal-fired power stations and steel mills with many more on the drawing board. The city of Shunning alone now pours out some 200,000 metric tons of sulfur dioxides a year, equivalent to a quarter of Japan's entire annual output, and the national emission rate seems certain to exceed that of the United States within a decade. More than 40% of China is believed to be affected by acid rain, and Chinese sulfur dioxide emissions have already had a serious impact on southern Japan. Not confined to China, the growing need for electrical power throughout Asia, especially in the so-called Tiger Economies of East Asia, has produced an increase in coal consumption of 5.5% a year in the last decade. The region now accounts for more than a third of the annual global output of sulfur dioxide gases and has already begun to suffer some of the pollution problems that presently confront China and Japan.[76]

Industrial emission controls introduced by most developed nations in recent years have helped to alleviate air pollution to some degree, but so long as fossil fuels remain a primary source of industrial energy the habitat destruction caused by acid rain will continue to be a major contributor to the environmental deterioration now occurring all around the planet.

The Final Equation

Every effort we have made to maximize primary production has led unavoidably to a greater expenditure of our real capital assets—clean air and water, fertile soil, and a diverse and supportive biota. The more we extract from the earth and sea the greater the environmental debt we incur, regardless of the techniques employed. Local damage may be minimized, even repaired in some cases, but only by dispersing the ecological debt over a wider area or over a longer term so that the local impact appears negligible. For example, if you reduce the intensity of farming in one area to allow the land to recover, food replacements must then be found from other sources. The increased yields of the Green Revolution might have postponed the food crisis that loomed before us in the 1960s, but they accelerated the degradation of the global environment and locked us into an even larger crisis in the future.

Whatever is taken from the earth or its incoming solar energy must be paid for somehow, somewhere, sometime. Each technological solution to our mounting ecological problems eventually exacts a commensurate fee, if not here and now, then elsewhere or later. It means that in the final analysis there is no such thing as a technological solution, there is only greater environmental debt. Consequently, the environmental impact of human beings depends absolutely on just three factors: the size of the population, its per capita level of activity, and the level of technology it employs. Biologist Paul Ehrlich and physicist John Holdren of Stanford University expressed this rigid relationship as an equation:

$$\text{Impact (I)} = \text{Population (P)} \times \text{Activity (A)} \times \text{Technology (T)},$$
$$\text{or } I = P \, A \, T.[77]$$

According to every scientific measurement yet devised, the biological health of the biosphere is currently in serious decline due to the impact of human beings. The only way to reduce that disastrous level of impact is to reduce the contributing factors in the equation. And there lies the dilemma. We will not voluntarily reduce our population, and we will not voluntarily reduce either our individual level of activity or the level of technology we employ, so the $I = P \, A \, T$ equation is, in effect, at its minimum value right now. But by the year 2050 the global human population, now almost 6 billion, will have grown by 20% to around 7.2 billion—at the very least.[78] To maintain even the present level of human impact on the planet under those circumstances, the $A \times T$ component would have to be reduced by a similar percentage (20%). But as we have seen, such a reduction is utterly beyond us. Our primary energy consumption quadrupled in the four decades between 1950 and 1990, and given the continued explosion of technology and consumerism, our accelerating energy demands show no sign of diminishing in the near future, in which case the next five decades will see us multiply our T factor by 2.5 at the very least.[79] Therefore, with only the A factor remaining constant, we are left with the dismal prospect that by 2050 the impact equation—at its very best—will look like this:

$$1.2P \times A \times 2.5T = 3 \times I.$$

And if the present impact equals serious global damage, then three times the present impact surely equals global catastrophe. The tide of life has finally turned against us.

Three obvious questions remain. How could we have fallen into this ancient and obvious population trap? Is there blame to be laid for the environmental damage we have caused? And where do we go from here?

Throughout the rest of this book I shall attempt to show that despite charges of greed, negligence, and malicious damage that could reasonably be leveled against our species, the mountain of circumstantial evidence against us is nevertheless thoroughly misleading. It is misleading because it is entirely based on the false assumption that we are primarily rational beings and not only have the capacity to modify our behavior but bear the responsibility to do so.

We are so accustomed to viewing human behavior in this light, however, that it requires considerable effort to unburden ourselves of such bias. To avoid falling back into the anthropocentric trap and slipping yet again into the old and comforting thought patterns that were instilled in us from birth—perhaps even wired into us before birth—we need to find a new and unfamiliar starting point. The best starting point is, of course, the beginning; and the beginning, as far we are concerned, is that single fertilized egg that attached itself to the wall of our mother's womb so many years ago. So let's now delve back into our own past and try to funnel our thoughts down a mental microscope that is focused on that newly fertilized cell, the cell that will become the organic entity known to each of us as "me." Then let's gradually tweak up the power of the microscope until we are able to penetrate the cell's tough outer walls. Once through them, we can then push our way through the cell's cytoplasmic matrix, past the many bacterial structures that crowd its bustling interior, until we come eventually to the thin membrane sac that encloses the cell nucleus. Here, at last, is our starting point, for only here can we confront the writhing coils of genetic material that ultimately determine the nature of all earthly existence.

PART II

Origins

Chapter 3

Our Genetic Origins

The Unity of Life

I f we were able to peer directly into the nucleus of the fertilized egg from which we grew, we would find ourselves confronted by forty-six tubular bundles of tightly coiled genetic material, all of them linked into pairs. These twenty-three pairs contain the coded instructions that make us what we are. Although this massive code message represents a blueprint for life that is utterly unique and distinct from that of any other person, merely by looking at a strand of it we could not even be sure that it belonged to an animal, let alone a human being. In fact, if it were not for certain peculiarities in its packaging we might very easily mistake our own genetic material for that of an oak tree, a mushroom, or a microbe. The reason for this indistinguishable quality is that the molecular constituents of all genetic material on earth are identical. In other words, although the astonishing diversity of life that currently inhabits this fertile planet gives little outward indication of a common origin, at a molecular level the evidence is overwhelming.

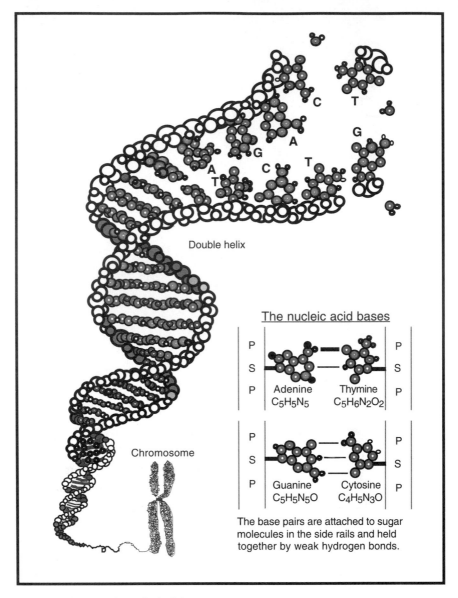

Figure 12. DNA—the code for life.

This primary evidence is spelled out in just four letters: A, C, G and T. These letters stand for the four main building blocks of the chemical code that regulates the life of every living organism. They are the nucleotide bases adenine, cytosine, guanine, and thymine, which bond together in only two combinations, adenine with thymine (A-T) and cytosine with guanine (C-G), to form the locking points of a flexible, zipperlike code that programs the life of its cell. Known as deoxyribonucleic acid, or DNA, this chemical zipper is the unseen dictator that controls every living organism, telling each precisely how and when to grow, reproduce, and die (see figure 12).

Twisted into a spiral and compressed into tight coils, this zipper is too small to be seen by the naked eye. Uncoiled however, the DNA of some species is more than one meter (3 feet 3 inches) long. Human DNA, incorporating some 3 billion base pairs, is one of these. About 97% of the DNA is never expressed (never translated into proteins) in us. The other 3% (about 3 centimeters, or 1.2 inches) represents an operational code of some 90 million individual "letters" spelling out a complete set of instructions for our growth, reproduction, and death. The total complement of genetic material within an organism, or within each of its cells (depending on the context of the reference), is known as its genome.

Besides defining the structure of an organism, the genome also incorporates instructions that provide a pattern of behavior designed to maximize the organism's chances of survival in a particular environment. Without precise operational instructions, even the best structural equipment in the world would be useless.

Replicating Our Genes

Most crucial of all, however, is the ability of genetic material to persist through geological time by accurately replicating itself in successive generations. The characteristic that enables DNA to maintain this continuity is its ability to unzip itself and construct new molecular "teeth" to lock onto the open side of each unzipped strand. Since there are no alternative patterns of linkage—A links only with T and C with G—each side of the DNA zipper forms a precise template for the construction of a complementary set of zipper teeth running in the opposite direction. This ability to rebuild each side of the DNA zipper to form two identical strands is the key to all earthly life.

At times that are appropriate for each genome, sections of the double helix are unzipped and turned on in response to codified cues embedded in what is called a promoter segment. The unzipped section is then read, three letters at a time, and copied (a process known as transcription), and the copy is sent to the cell's factory site outside the nucleus for translation into a specific sequence of amino acids. The three-letter "words" (known as codons) that dictate these instructions encode only two kinds of statement: an instruction to the factory to incorporate a particular amino acid into the production run and an instruction to cease production.[1] The assembled chains of amino acids produced during this process constitute the proteins that dictate the growth, maintenance, and behavior of the organism. Groups of codons that are read together are thereby the ultimate determinants of what the organism will look like and how it will behave throughout its life. These codon groups therefore represent the basic units of heredity. We know them as genes. At the beginning of their lives at least, all cells except eggs and sperm possess a full copy of all genes. (Eggs and sperm, the sex cells, contain only unpaired chromosomes and constitute half, or a haploid version, of the genome.)

Since the entire chemical vocabulary expressed by all genes consists of only twenty amino acids, the astonishing diversity of species produced by evolution therefore hinges on its ability to ring an infinite number of minor changes in the sequencing of those twenty amino acid instructions while maintaining the species' general viability. By this means, evolution has already produced during the past 4 billion years somewhere between 5 billion and 50 billion species—of which about 99.9% are now extinct. Although this astonishing diversity might at first glance suggest that evolution is peculiarly devoted to producing genetic variation, nothing could be further from the truth. We have the brevity of our lives to blame for this illusion, since it presents us with a compressed view of time that savagely distorts our perception of the evolutionary process. In fact, genetic variation occurs at a snail's pace and DNA represents, above all, stability.

Genetic Fidelity

Errors made during transcription and replication of genetic material are virtually nonexistent by human standards, not as a result of some Grand Design but as an inadvertent and unavoidable consequence of the laws of chemistry that govern it. There are even mechanisms in place to correct the rare mistakes that do occur, for although many mistakes have no identifiable effect at all, almost all others are lethal. "Fortuitous errors" are rare indeed. Considering the massive size of the human genome, the degree of fidelity achieved during the replication process is breathtaking. Especially when we remember that each cell must replace itself many times throughout a human lifetime and such replications are occurring in all parts of our bodies every second of our lives.

About 5,000 errors appear daily in the genome of each cell of the human body as part of the normal processes of decay. Almost all of these faulty "letters" are immediately replaced by built-in repair mechanisms, enabling the information encoded by a particular gene to survive virtually unchanged for millions of years, and in some cases for hundreds of millions of years.[2] As British evolutionist Richard Dawkins puts it, the central truth of life on earth is that living organisms exist for the benefit of DNA rather than the other way around. Each individual organism should be seen as a temporary vehicle in which DNA messages spend a tiny fraction of their geological lifetimes.

> Individuals are not stable things, they are fleeting. Chromosomes too are shuffled into oblivion, like hands of cards soon after they are dealt. But the cards themselves survive the shuffling. The cards are the genes. The genes are not destroyed by crossing-over, they merely change partners and march on. That is their business. They are the replicators and we are their survival machines. When we have served our purpose we are cast aside. But genes are denizens of geological time: genes are forever.[3]

It matters little whether you are a bacterium, a bristlecone pine, or a human being, as far as your genes are concerned "you" are merely a heavily armored vehicle specifically designed to ferry your precious fragment of genetic information through time and space until it can be safely transferred into a younger, stronger—and with luck—even better vehicle. Since genes are long lived and we are ephemeral, no other conclusion is available to us. In Richard Dawkins words,

> Now they swarm in huge colonies, safe inside gigantic lumbering robots, sealed off from the outside world, communicating with it by tortuous indirect routes, manipulating it by remote control. They are in you and in me; they created us, body and mind; and their preservation is the ultimate rationale for our existence. . . . We are their survival machines.[4]

In the natural world any genes that produce either inferior survival vehicles or disadvantageous behavior are soon weeded out by their high failure rate during field trials in the unforgiving environment. Conversely, a particular genetic heritage that enhances the viability of its survival vehicle is more likely to leave descendants than one that doesn't. In this endeavor each gene is also in continuous competition with others of its kind. Richard Dawkins again:

> It is its potential immortality that makes a gene a good candidate as the basic unit of selection. . . . Any gene that behaves in such a way as to increase its own survival chances in the gene pool at the expense of its alleles will, by definition . . . tend to survive. The gene is the basic unit of selfishness.[5]

I should explain here that an allele is an alternative form of a particular gene. It may be present in the same individual on the other chromosome of the pair, and any particular gene may have many alleles scattered throughout the population. It is almost certainly the allelic presence of our genes in others that generates our genetic imperative to behave altruistically toward them whenever their welfare is threatened.

Parental Sacrifices

Four billion years of field testing and rigorous Darwinian selection have gone into producing the earth's current batch of genetic formulas. The result is a biota armed to the teeth with structural and behavioral safeguards that are, in this evolutionary sense, specifically designed to protect the genes from the slings and arrows of a fickle environment. To this end, many species are coerced into providing extended altruistic nurture for their progeny, as their own genes attempt to protect the descendant alleles that will carry their genetic heritage into the future. We humans, for example, feel driven to feed, educate, and protect our children into their teenage years and beyond, often at great personal cost. Most of us find that our feelings of parental duty linger long after our children have launched their allelic genes to produce our grandchildren.

Like other animals, however, we are unconscious of the real origins of this altruistic behavior, and we are genetically seduced into the parental role by far more powerful forces than reason could provide, forces that manifest themselves to us in the guise of overwhelming parental love and irresistible feelings of responsibility. So powerful is this genetic urge in most of us that it spills over to include children in general—since all humans contain many genes that are allelic and recognizable to ours, and those in children are not yet directly competitive.

Gorilla Mother

One of the most intriguing examples of this larger parental imperative was displayed at Chicago's Brookfield Zoo on 16 August 1996, when a small boy climbed a barrier-planter and its railings and tumbled more than 6 meters (some 20 feet) down a ravine into the gorilla habitat. Crowd and staff watched in horror as seven lowland gorillas stopped what they were doing and an eight-year-old female, carrying her seventeen-month-old baby on her back, approached the body of the unconscious boy. She picked him up, cradled him in her arms, carried him across to an access door, and quietly waited there for the keepers to come and take over the parenting role that she had so readily assumed.[6] That the gorilla, Binti Jua, had been hand-reared by zoo staff does not appear to be particularly significant because similar gorilla behavior toward humans has been documented in two other instances. One occurred at Jersey Zoo in the Channel Islands and another at Tama Zoo just outside Tokyo. These examples of interspecies compassion also serve to illustrate that gorilla genes recognize themselves in us—perhaps even more readily than ours recognize themselves in gorillas.

While accepting that Binti Jua's behavior might have illustrated a primal drive common to both species, many people consider that the flexibility and complexity of human behavior is well beyond the productive capacity of genetic material. It should be remembered, however, that there are roughly 100,000 genes in human DNA, and some of our genes are very large indeed, incorporating up to 2.5 million individual code "letters," or more than 800,000 codon "words." The literary equivalent of such a gene would be eight books the size of this one. We seem to be extraordinarily well equipped for complex behavior. Besides, we now know that particular genes do not code for particular behavior, as was once thought. Instead, it seems that large gangs of genes, opportunistically grouped into temporary coalitions, continually collaborate and compete with each other in their efforts to influence our lives.[7] In anthropomorphic terms, this means that we live, not according to the tyrannical edicts of a multitude of minor genetic despots, but according to a set of shrewdly flexible guidelines thrashed out over millions of years by a host of experienced gene committees operating secretly within the stable parliament of the human genome.

Genes have one permanent, implacable enemy: the environment. It keeps changing the rules of the survival game; evolution, like Lewis Carroll's Red Queen, must run to keep up. The wider the spectrum of viable variations a species can present to a changing environment, the greater the chance of finding at least one variant that is suited to the new conditions. Unfortunately "good" genes are painfully slow to evolve and genetically expensive to produce—so expensive that evolution found an easier, cheaper way of producing new species.

DNA: The Traveling Circus

Whereas genes often took millions of years to perfect, their immediate product, protein, was cheap to manufacture, spectacularly varied in form and function, and readily disposable. By making relatively minor modifications to protein production runs, evolution was able to dramatically change the look and performance of a species while leaving its precious, time-tested genome largely intact. This strategy enabled evolution to become a traveling circus in which a relatively small cast of genes could keep changing the show merely by varying the costume and makeup. The

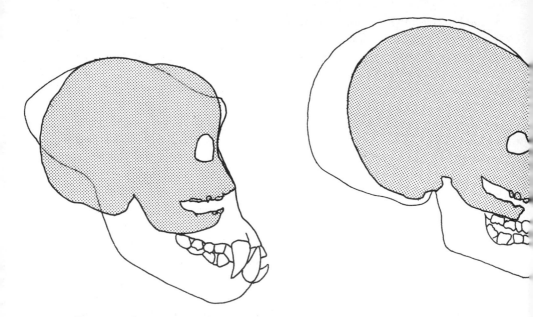

Figure 13. Comparison of infant and adult skulls in chimps (left) and humans (right).

fur, the feathers, or the foliage might differ dramatically from species to species within the one family, but beneath the surface it is largely the same old crew of genes strutting their stuff.

Should the genetic microswitches that control the growth pattern for a particular part of the body become reset, the descendant form may display a marked difference from its antecedents. The horse is a striking example. It still begins its life with five toes programmed for each foot, just as mammals always have, yet almost all horses are born with only one effective toe on the end of each leg. This toe is oversized and tipped with a massive nail that forms the hoof. While the feet of the modern horse may look dramatically different from those of its twenty-toed primogenitor, significant structural alterations like this represent only minor modifications at the genetic level.[8] They are entirely governed by the multitude of microswitches embedded in the side rails of the species' DNA. These switches turn on or shut down protein production in each cell at times that are appropriate to the cell's role in the organism. In the DNA of a modern horse, the switches that determine the growth of four of its toes now ensure that they stop growing very early in its fetal development, allowing the remaining digit—number three—to continue growing until both toe and toenail are disproportionately large. We know that is what happens because a recent ancestor of the horse had an intermediate arrangement (four toes in front and three behind), and even today horses are occasionally born with one or two additional toes on their feet. In fact, all modern horses still retain vestiges of their second and fourth digits in the form of two bony splints that lie concealed against the tibia, high above the hoof. The first and fifth digits, however, fail to develop at all.

Similarly, if we compare a human baby's skull, in profile, with that of a baby chimp, the similarity is unmistakable. Yet within a few years that basic similarity will have vanished (see figure 13). The lower part of the chimp's face changes dramatically as it matures, whereas the facial bones of an adult human retain much of their baby-chimp architecture.

The retention of juvenile features in the adult, a phenomenon known as paedomorphosis, appears to have provided many ancestral animal species with the means of escaping from the evolutionary cul-de-sac of their morphological specialization. The particular form of paedomorphosis that molded the human skull and its contents into their present shapes is known as neoteny.

Human Origins

Genetic changes like these accumulate in a ratchetlike process that has been compared to the ticking over of a car's odometer. For most of the time, an organism's appearance does not alter much. Occasionally, however, it takes only one more minor click to alter it drastically. For example, to change an odometer reading from 99,999 to 100,000 represents the addition of only one unit. And once the genetic odometer has ticked over, as in the car, there is no going back. Our genetic switches have been reset many times in this fashion during the past 4 million years of hominid development. The changes they produced were, of course, not restricted to the face but were expressed throughout the body and resulted in a significant shift in the type of animal being presented to the Darwinian processes of natural and sexual selection. Where the changing climate produced regional differences in the hominids' habitat, different suites of genetic variations would have been favored in different regions, producing localized subspecies. As environmental pressure increased with the onset of the ice age, this speciation process seems to have accelerated, and by about 2.5 million years ago the hominid branch of the primate family had subdivided into at least three, and perhaps even five or more, separate species.[9]

Changes in one of these groups produced a slightly larger animal with a disproportionately large brain and a skull that incorporated several unmistakably human characteristics absent in other hominids of that period. Several 2-million-year-old fragments of such a skull, known as

Skull: KNM-ER 1470. The skull known as ER 1470 is generally recognized as one of the oldest human relics ever found. A cast of its cranial cavity, bearing the faint signs of an incipient Broca's bulge, tends to confirm this classification. Paleoanthropologist Richard Leakey and his team unearthed the skull fragments at Koobi Fora in Kenya in 1972. (Cast courtesy of the Cleveland Museum of Natural History.)

KNM-ER 1470 (the catalog label assigned to it by the Kenya National Museum), were found in 1972 at Koobi Fora, on the eastern shores of Lake Turkana in Kenya. Since the skull fragments were found in association with some of the world's oldest manufactured stone tools, the original owner was classified under the title *Homo habilis*, or "handy man." It was not until an endocast (an interior cast of the brain casing) was taken by Ralph Holloway of Columbia University more than a decade later that researchers detected on the left front of the endocast the faint impression of a thoroughly modern pattern of sulcal folds in the region of Broca's area.[10] The neurological key to language had already slipped into the lock. Human evolution had begun.

I am unhappily aware that by adhering to traditional anthropological classifications and using such labels as *H. habilis*, *H. erectus*, and *H. sapiens*, I am propagating the myth that our modern species represents a considerable genetic divergence from our evolutionary siblings the chimpanzees *Pan paniscus* and *Pan troglodytes*. Molecular biology not only suggests that we barely deserve a generic label of our own but also shows that the unequivocal species distinctions we have tried to apply to human remains is unrealistic and misleading. According to many molecular taxonomists, if the rules of distinction customarily applied to other animals were extended with equal rigor to our kind, we should be classified as *Pan sapiens*, a third species of chimpanzee. However, human "honor" has always had difficulty with this apparent demotion, so to minimize confusion I shall uphold tradition in this case.

Homo habilis

Two different dating techniques have determined that the owner of skull ER 1470 died almost 2 million years ago. Despite its hominid classification, it represents an animal that was not so very different from other so-called missing links that have been found in recent years. Essentially it was a smallish apelike hominid, 100 to 130 centimeters (40 to 51 inches) tall, habitually erect and bipedal, with shortish legs, and relatively long, chimplike arms, but its skull showed significant divergence from its ancestral line. The cranial cavity was considerably larger than that of the other hominids roaming East Africa at that time, and its brain would have weighed approximately 750 grams, as opposed to a mean brain weight of only about 450 grams for all other great apes, including ER 1470's immediate ancestors, the gracile australopithecines.[11]

Since brain size is primarily related to body size, it is misleading to use brain volume as a guide to intelligence. A better guide is to compare the brain-to-body ratio of the species in question to that of other animals of

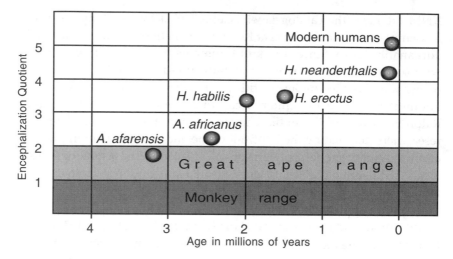

Figure 14. Primate encephalization quotients. The graph shows how the encephalization quotient has changed over the past 3 million years. The early australopithecine hominids were well within the ape range, and although the increase that occurred with *Homo habilis* was significant, the major expansion did not take place until relatively recently with the appearance of *H. sapiens*. Some Neanderthals had larger brains than modern humans but their greater body mass sets their EQ below ours. (Modified from *New Scientist*, 14 October 1989.)

equivalent body mass. This gives us what is known as an encephalization quotient, or EQ. If we take the EQ of the other great apes as a baseline and arbitrarily set this base between 1 and 2, then you and I have an EQ of slightly more than 5 (see figure 14). According to this scale *Homo habilis*, with an EQ around 3.3, lay halfway between ourselves and the other modern apes.[12] Since the thinking capacity of gorillas and chimps (EQ 1–2) roughly matches that of a 2–3 year old child, it seems probable then that *H. habilis* possessed the perceptive powers and reasoning capacity of a modern 6–8 year old. If that was the case then they clearly had a head start on the competition.

It is generally conceded that the presence of an incipient Broca's area in the brain of ER 1470 is not evidence that *H. habilis* possessed language as we understand it. However, it may well have afforded them the basic neural equipment for it, thereby offering what amounts to a quantum leap in their capacity for reasoned thought, technology, and communication. The evidence for this is subtle and circumstantial, but unequivocal. The endocast taken from the reconstructed skull of ER 1470 shows pronounced asymmetry between the right and left hemispheres in what is known as a left-rear, right-front petalia pattern. This asymmetry suggests that *H. habilis*, like modern humans, was predominantly right-handed—another argument in favor of an incipient language capacity, since speech organs and right-hand movements are controlled by adjacent areas in the left frontal lobe. Their right-handedness is born out also by the way in which they chipped the cutting edges on their stone tools. Similarly, 80% of the ocher hand stencils left by ice-age hunter-gatherers on the walls of European caves show left hands. A reverse petalia (left front, right rear) is common with left-handers, and more common among women.[13]

With the possible exception of the orangutan, the great apes display relatively little brain petalia and only limited signs of right- or left-handedness. However, the behavior of many other animals suggests that some degree of asymmetry is common, even in animals as diverse as whales, rats, and birds. It is particularly evident in birds that sing complex songs. A male canary's ability to advertise its genetic potential by its voice appears to be controlled almost exclusively by the left side of its brain. Just before the mating season each year, this region of the brain swells with a multitude of new neuronal connections, only to shrink back to its usual size at the season's end. This ability to develop new circuitry in response to use is known as brain plasticity.[14]

The Malleable Brain

The basic neuronal equipment for human life is wired into fetal brains at the astonishing rate of 250,000 new neurons a minute during most of the nine-month gestation period, and all of this basic circuitry is laid down in accordance with instructions encoded in our DNA.[15] As soon as postnatal awareness begins, however, this basic wiring becomes selectively reinforced and expanded by our particular patterns of use. In the words of Carla J. Shatz, professor of neurobiology at the University of California, "Cells that fire together wire together."[16] By this means, frequently used circuits continuously extend, enabling hundreds of new connections to be made, to the point that the adult brain contains something like 100 billion neurons, all of them linked by at least 100 trillion nerve synapses. It has even been suggested that the connection alternatives inherent in this astronomical number of synapses are greater than the total number of atomic particles in the known universe.[17] We are wired for complexity.

The plasticity inherent in such a process allows individual experience to directly influence the brain's wiring layout, fine-tuning the cortex to a degree unmatched in later life. After about ten years, the brain gains no further neurons and vast numbers of little-used connections begin to atrophy and die. This is why we find it so much easier to learn basic skills during childhood. Language is the classic example. Children who have for various reasons been deprived of the opportunity to learn any kind of language in their early years find the task increasingly hard as they approach puberty, regardless of their intelligence. By the age of thirteen, the learning window closes, and the acquisition of complex language becomes impossible.[18]

Victor and Genie

Two well-documented cases attest to this phenomenon. The first was that of "Victor," the "wild boy" of Aveyron, France, who lived most of his twelve years totally without human contact in the woods near Aveyron. Captured in the late eighteenth century, he was given to a sympathetic university professor who tried for five years to teach him language. Victor learned to say only "Oh Dieu!" and "lait" (milk). Similarly "Genie," a tragically abused Californian girl, who had been imprisoned alone in a darkened room for almost twelve years by her deranged father, emerged from her ordeal with no knowledge of the outside world and not the faintest glimmering of language. During the whole of that time her father had used no words in her presence and had treated her like a dog, beating her and barking when she displeased him. When questioned, her mother said that Genie had been about twenty months old when her father had locked her away. Until that time she had developed normally and had begun to speak a few words. Having begun to talk during those crucial twenty months, a rudimentary circuitry for language seems to have been locked in, enabling her to acquire a limited vocabulary after her release. Significantly, she failed to master even the most elementary grammar, an asset normally acquired between the ages of two and four when the child progresses from two-word utterances to sentences. When daily lessons ceased, Genie's vocabulary shrank and her limited capacity to communicate soon slipped away. In the words of neurobiologist Jean-Pierre Changeux, "To learn is to stabilize preestablished synaptic combinations, and to *eliminate* the surplus." (The italics are his.)[19]

Since newborn babies show no sign of a selective enlargement of the brain in Broca's area, the enlargement of Broca's area that is discernible in adults almost certainly occurs in response to neuronal stimulation and dendritic growth engendered by the learning of language. Similarly, the presence of a discernible Broca's bulge in the ER 1470 endocast is arguably due to its owner's exercise of this language facility during the brain's highly plastic developmental period.

Common Language

The significance of the proximity and close liaison between Broca's area and the motor control region in humans seems to be that both are prima-

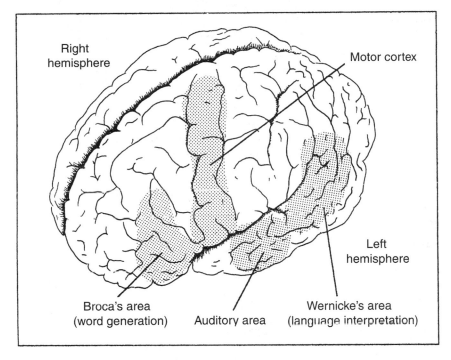

Figure 15. Areas of the brain associated with language.

rily concerned with sorting and sequencing previously filed information in order to produce a reasoned progression of vocalizations or hand movements. This is why speech is so often accompanied by gesticulation (see figure 15). Similarly, experiments in teaching sign language to deaf children of various ages have shown that they too display the patterns of ability, disability, and stage-by-stage development according to age that hearing children display. Not only do they both learn language with the same speed and degree of competence, but damage to Broca's area in a deaf child produces dysfunctional signing that closely matches the dysfunctional speech produced by similar damage to the brain of an otherwise normal child.[20]

Although this close partnership between Broca's area and the motor control region seems to be especially productive in humans, there is every reason to believe that human language and modern technology merely represent the mature fruits of a neuronal marriage that probably dates back many millions of years. The ability of all the great apes to communicate with humans in very simple terms using sign language—either American Sign Language or pictorial symbols on a computer—suggests that close links were already well established between Broca's area and the motor control region by the time apes appeared some 15 million years ago. It also shows us that ape communication and human language represent different points on a language continuum and that our ability to construct and interpret language is as innate as our ability to interpret the data that come from our senses as we see, hear, smell, and feel our way through life.

Since language is therefore primarily a product of DNA rather than intellect, the appearance of Broca's area in *Homo habilis* more than 2 million years ago represents a significant milestone in the slow divergence of our kind from the rest of the primates. It would alter forever the nature of the relationship between this animal and its environment and take it down a wholly new and untrodden evolutionary path. Surrounded as we are by environmental problems that we have brought on ourselves in traveling that path, we might well look back somewhat ruefully as we begin to understand that our successful deviation from the main highway of mammal behavior was largely climate driven in the first place.

Increased concentrations of the isotope oxygen-18 in both Atlantic and Pacific Ocean sediments and of carbonates on sub-Antarctic seabeds confirm paleobotanical evidence from the Omo River in Ethiopia that between 3.3 and 2.4 million years ago there was a significant temperature decline around the world.[21] Antarctica's ice cap, which had long ago reached sea level, began to expand out over the Southern Ocean, and the northern polar ice cap formed. Continental ice sheets accumulated around its fringes, and the Great Lakes region of North America began to disappear beneath a mile of ice. The oncoming ice age, which had been tightening its screws on the planet for some 35 million years, was about to launch a series of glaciations that would transform plant and animal habitats beyond recognition. And it would send the processes of evolution and speciation into high gear.

Homo erectus

As temperatures fell and climates became more seasonal and unreliable, large areas of East African forest were replaced by open woodland and grassland. A hominid species that could bring better planning and communication skills as well as superior weaponry to the business of protecting and feeding itself on those hazardous open plains would have had a decisive advantage in the struggle for survival. And somewhere around 2 million years ago the grim process of environmental selection sifted through the *Homo habilis* variants and came up with a slightly modified form that worked very well indeed. This descendant, *Homo erectus,* still bore many of the hallmarks of its primate ancestry, but its flared and flattened pelvic structure, flatter face, large asymmetrical brain, and pronounced Broca's bulge, proclaimed its further divergence from the rest of the hominid line. Wear marks on the teeth of *H. erectus* suggest that their diet included meat on a regular basis. Unlike their ancestors, *H. habilis,* who appear to have supplemented their 80% vegetarian diet largely by scavenging from the carcasses left by other predators, *H. erectus* was clearly a hunter as well as a gatherer.[22]

Judging by the close relationship between diet and EQ in other primates, this climate-driven dietary change in our ancestors was the primary trigger for the development of the most crucial evolutionary asset ever to appear in our genus—an oversized brain. Comparative studies of monkeys that are closely related and live in similar habitats shows a curious divergence in their EQ, a divergence that is essentially diet-related, according to Katharine Milton, professor of anthropology at the University of California–Berkeley. Milton found that the black-handed spider monkey (*Ateles geoffroyi*), a dedicated fruit eater, boasts almost twice the brain mass of its leaf-eating cousin, the mantled howler (*Alouatta palliata*), despite their being of similar size, having simple, unsacculated stomachs, and gathering their food in the same forests on the island of Barro Colorado in the Republic of Panama.[23]

This diet-related EQ dichotomy reappears throughout the animal world but is especially marked among primates, probably because monkeys who live almost entirely on leaves, and therefore do not have to travel very far to find their food, do not require large brains to live successfully. Fruit eaters, on the other hand, must be able to compile detailed mental maps of their home range and correlate those maps with an accurate, seasonally adjustable mental calendar so that they can most efficiently harvest their ephemeral and scattered food resources. Such fussy eaters not only must be able to converge on their favorite fruit trees by the shortest, most economical route, but must also undertake long journeys to new fruit crops only when the pickings are likely to be worth the energy expended in reaching them. In other words, they must be able to plan ahead to some extent and be willing to take well-judged risks. Both chimpanzees and orangutans exemplify the fruit-eater rule, and therefore in all probability so did our chimplike common ancestor. As a result, all developed relatively larger brains than their massive, leaf-eating cousin, the gorilla.

Building Better Brains

When our ancestors were gradually forced to alter their diet, switching first to seeds, nuts, and roots and finally to meat, the difficulty involved in finding those foods would have pushed them toward a further upgrade of their mental equipment, eventually resulting in a degree of brain enlargement comparable to that which now distinguishes the fruit-eating spider monkeys from their leaf-eating cousins, the howlers. During the last 2 million years, as a result of a dramatic shift in diet, the human brain has almost doubled in volume. Even more significantly, in just 3 million years the surface area of the human cerebral cortex has quadrupled, expanding from 500 square centimeters, the area possessed by chimpanzees, to an average of about 2,000 square centimeters, a rate of change unprecedented in the evolution of other species.[24]

Another factor helped to lock us into that particular mode of evolutionary development. Fruit eaters and leaf eaters choose to eat as they do largely because of a difference in gut design and because of the relative speed and efficiency with which their slightly different intestinal tracts digest their preferred food. Like all fruit eaters, spider monkeys process their food quickly but extract a smaller percentage of its available energy than do those that eat more fibrous vegetation. Not only must they therefore eat very selectively, but they must also eat very often.[25]

When the changing climate robbed early hominids of the continuous supply of high-energy fruits that their ancestors had enjoyed, they would have found a diet of coarse woodland leaves a thoroughly unsuitable alternative. They did not have the stomach for it. Stomachs, it seems, are tricky things for evolution to modify: brains are far easier. So the best option left to our hard-pressed ancestors was to increase their protein intake by spending more time hunting live prey—and to develop larger, more efficient brains to cope with the complex problems that this new and risky lifestyle entailed.[26]

Large brains are expensive accessories, however—materially expensive to build and energy-expensive to operate and maintain—and with hominid backs already firmly set against the evolutionary wall, the development of an oversized brain would seem to be a high-risk venture indeed. For example, a modern human brain represents only about 2% of the body's weight but requires about 16% of the blood supply, about ten times as much blood per unit mass as muscle tissue requires.[27] Nevertheless, our very survival indicates that the gains must have been both disproportionate and immediate for such brain development to have proceeded beyond its *Homo habilis* prototype.

When *H. erectus* first appeared, it still shared the hominid stage with at least one, and possibly three, other species, all of them vegetarians. By 1 million years ago, however, this large-brained omnivore not only was the last hominid species left on earth but had succeeded in dispersing with explosive speed halfway around the world. So, if we are to unravel the secrets of humanity's successes and failures and account for our own behavior, both "good" and "bad," it is to our most immediate ancestor, *Homo erectus*, that we must look. As their direct descendants we must face the fact that it is almost entirely their genes that characterize modern human DNA, and it is their genetic imperatives that have shaped human culture in the past and will continue to do so in the future.

Late Developers

We know relatively little about the lifestyle of these hunter-gatherers, but careful analysis of the skeletal debris they left behind has revealed three subtle but crucial differences from their ancestral family, the australopithecines. Most important was the change in their maturation rate: their developmental processes were significantly slower. Consequently, they were born in a relatively premature state and took longer to grow to full size and attain sexual maturity. But this apparent handicap concealed a crucial advantage: the modern human brain emerges from the womb at 24% of its final weight and only partly developed. It then continues expanding in explosive bursts for a further ten years. By contrast, a newborn chimpanzee brain is about 60% of its final weight, and its postnatal growth ceases within a year or so.[28] Even a small increase in human brain plasticity 2 million years ago would have opened a crucial learning window for *H. erectus*, giving them a head start on other hominids. Their slower maturation rate compensated for this educational time-out by lengthening their life span.

A clue to all these changes lies in hominid teeth. Teeth are the most durable part of a mammal's body, and they fossilize well. Meanwhile they reveal much about their owners' nature and lifestyle by their structure and position in the jaw, their chemical composition, and the kind of wear and tear they display on their surfaces. Teeth also contain incremental growth patterns, known as *striae of Retzius,* that are laid down every seven or eight days as a result of cycles in the secretion of enamel. They are rather like the growth rings in trees, except that they appear to be biological constants in all primates, irrespective of body size, enabling the relative age of the individual at the time of death to be determined very precisely. Analysis of these enamel layers shows that *Homo erectus* babies were born in a relatively underdeveloped state compared to the babies of other primates; they also grew more slowly and took much longer to reach physical and sexual maturity.[29] These modifications combined to produce slow-maturing, slender, baby-faced adults that lacked coarse body hair and long fighting teeth. In fact, they retained most of their juvenile ape features and body proportions, including a relatively large head and a disproportionately large brain. In other words, they were creatures very like ourselves.[30]

In addition, *H. erectus* possessed a greatly extended vocal flexibility due to a rearrangement of the palate and larynx. Even today human babies are born with the old primate structures intact, allowing them to drink and breathe at the same time. Near the end of their second year of life, however, the larynx descends to a position in the throat that makes this impossible but produces a greatly expanded pharyngeal chamber above the vocal chords. Sound made by the larynx can then be modified to a far greater degree than is possible in infant humans, chimps, or indeed any other mammals. This represents the key structural modification that allowed human speech to develop.[31]

By the time *H. erectus* appeared, this pattern of development seems already to have occurred to some extent, and although the degree of laryngeal modification was moderate, it would certainly have permitted them to develop a much more complex system of vocal communication than their australopithecine ancestors or their ape cousins.[32] Used in conjunction with hand signs, this increased verbal communication would have given *H. erectus* a distinct advantage over other hominids.

The Price of Neoteny

As with all specialization there was a built-in fee, however, and several of the modifications that *Homo erectus* displayed would have presented the species with very serious problems. Not only were their incompetent offspring more vulnerable to death by accident or predation for a much longer period, but also their prolonged childhood represented a major handicap to parents who had to feed and protect them. There were physical losses also. Although the process of neotenization brought with it a much larger, more adaptable brain, it meant the loss of much more than just a coat of coarse protective hair—it cost them most of the strength, speed, and agility that other apes absolutely depend on for survival. A male orangutan, for example, has about the same body mass as a small sumo wrestler but is roughly seven times stronger than the average man and much more agile. (Even so, male orangutans avoid traveling on the ground wherever possible.) In fact, the only physical assets of real value that our ancestors retained in the genetic trade-off were a pair of very flexible shoulder joints, manipulative hands, and good stereoscopic vision.

On the face of it, then, their losses amounted to a litany of deconstruction that might well have proved fatal for the species in that hazardous and unstable environment. That it didn't indicates that hidden benefits must have outweighed those spectacular losses. Clearly, the decisive advantages must have been behavioral, most of which undoubtedly lay in the sophisticated data-handling facility that had begun to develop in the expanded frontal lobes of their outsized brains. Certainly this new facility did not amount precisely to the addition of a new sense per se, but in operation it bore an uncanny resemblance to the mammalian sense of smell in that it enabled its owners to sift quickly through a deluge of abstract data and act on the few scraps of information that were relevant to them, very much the way other mammals sort through a barrage of olfactory stimuli.

With the development of this remarkable "sense" of information, *Homo erectus* regained some of the evolutionary ground its tree-dwelling ancestors had lost tens of millions of years earlier when they exchanged their long sensitive noses for the primate's flattish face and forward-facing eyes. In effect, *H. erectus*'s enlarging brain not only reequipped them with what amounted to a novel version of that ancestral mammalian nose, but it did so without disturbing their good stereoscopic vision (which the reinstatement of a longer and therefore more sensitive nose would have done by disrupting the field of vision). This informational "nose" was to prove especially valuable, defining and solving *H. erectus*'s technical and tactical problems with increasing efficiency during the next 2 million years. That this sophisticated neuronal facility should have evolved so early and so swiftly in our human ancestors might seem remarkable—until you look at the evolutionary origins of the brain structures that nurtured it.

The Nose Knows

All mammals originated from a few nocturnal, shrewlike creatures and exploded into their modern diversity after dinosaurs became extinct 65 million years ago. Like their ancestors, modern shrews tend also to be nocturnal and depend absolutely on their highly developed sense of smell. Consequently, most of the forebrain in these animals is still devoted to their sense of smell and represents a highly sophisticated olfactory receiving station that has changed little in 100 million years. The primate forebrain is a greatly enlarged version of that same olfactory analysis center, but its function has been hijacked and transformed.[33] We can gain an idea of how and when these momentous changes occurred by looking at the modern products of the evolutionary progression that links us to our shrewish ancestors. Small insectivores that resemble the ancestors of primates still devote almost 9% of their brains to odor analysis; for prosimians the figure is about 1.8%; for monkeys it comes down to 0.15%, and for great apes, 0.07%. Despite massive enlargement of the frontal lobes, only about 0.01% of the human brain is still concerned with odor analysis.[34]

Central to the human forebrain is a region known as the prefrontal cortex. Current research suggests that this area is solely responsible for executive functions such as strategic planning and the allocation of cognitive

Long-nosed bandicoot (*Perameles nasuta*). Australia's long-nosed bandicoots use their shrewlike noses and sophisticated olfactory abilities for locating and digging out small burrowing invertebrates.

resources.[35] It allows us to juggle several ideas at once, comparing, discarding, selecting, and collating them to provide a reasonably rational decision-making system. In effect, we now deal with intellectual information and abstract ideas much as our shrewish ancestors would have dealt with olfactory information. Here lies the neuronal underpinning that set *H. erectus* on its meteoric path to the very top of the world's food chains, and eventually out into space. How delicious that all of this should rest in an evolutionary sense on a shrewish nose and forebrain—olfactory equipment so efficient that it still makes an appearance in my garden every night in the shape of a long-nosed bandicoot (*Perameles nasuta*).

As my bandicoot friend rummages through my unkempt garden each evening, sniffing out tasty subterranean creatures and sorting out the tangle of odors and pheromones that daily occupy its territory, so my prefrontal cortex continually rummages through tangles of intellectual and abstract data, sifting out the "juicy" bits, "tasting," sorting, and reacting to them. And no doubt our shrewish ancestors were doing the same among the dinosaur droppings when the comet hit 65 million years ago. Thanks to their sharp wits and highly developed sense of smell, they survived the global carnage of that time and lived on to give rise to a multitude of mammalian descendants. Plumped up and squeezed into a new shape, the shrewish forebrain has become our neuronal Swiss army knife, an all-purpose mental tool that enables us to devise the technologies and build the social structures that characterize modern civilization.

We unwittingly commemorate this momentous evolutionary link between intellect and smell every time we trot out the olfactory metaphors that stud modern human language. A proposition stinks; we get wind of a plan, follow the scent of an idea, get a whiff of scandal, smell a rat, develop a nose for news, and sniff out a good story. In fact, were it not for this evolutionary link and consequent operational similarity between mammalian odor analysis and the sorting of abstract data by the prefrontal cortex of the human forebrain, such references would appear to give a strange emphasis to a sense that no longer plays a major role in the daily survival of our species.

Living a Double Life

All mammals exist on two levels. Not only do they inhabit a physical environment made up of air, water, earth, and other life forms that they can see, touch, and taste, but they also inhabit an invisible landscape made up of smells, sounds, heat, and a complex pastiche of memories, fears, and expectations. It is primarily on the basis of such intangibles that they make the most important decisions of their lives—when and where to find food, or a mate, when to stand and fight for their territory, and when to run for their lives. We too inhabit both of those worlds, one tangible and visible, and the other invisible and abstract. As with other mammals, it is the invisible world that dictates most of our behavior, communicating with our instincts as directly and urgently as the odor trails in my unweeded garden communicate with my resident bandicoot. Prevailed upon by my informational "nose," I too make choices I cannot help making, given the prevailing landscape of information. Regardless of the careful reasoning I may marshal in support of my actions, the choices I make are ultimately animal choices. In fact, the only thing that truly distinguishes humans in the arena of animal behavior is our naive belief that our decision-making processes are primarily cortical and rational and therefore unlike those of all other animals. (We will explore this illusion at length in chapter 8).

Our *Homo erectus* ancestors would have been less troubled by our tendency to theorize in this fashion, and lacking both the language complexity and the mental capacity to explore the long-term consequences of alternative behavior, they would have obeyed their genetic imperatives with even less hesitation than we do. Their growing ability to recall and compare a wide variety of "filed" data, and to use this information to predict events and plan ahead, either to take advantage of or to escape the worst consequences of those events, would have tipped the evolutionary scales decisively in their favor. Here at last was a much smarter brand of hunter-gatherer, one less likely to blunder blindly into a waiting leopard's jaws or to starve to death when the rains failed. And here also was an imaginative, resourceful adventurer.

Hominid Dispersal

About 2 million years ago *Homo erectus* became the first hominid to set foot outside Africa, moving first into the Middle East and then dispersing swiftly through southern Asia all the way to the shores of the Pacific Ocean. This urge to travel may well have been triggered by their increasing efficiency as hunters. A large predator, hunting large prey, needs plenty of hunting territory; as the predator population outgrows its food resources, the hunters are continually forced to expand their frontiers and stake new territorial claims. Assuming only a modest 10-kilometer (6-mile) expansion per generation, *Homo erectus* could have covered the distance from East Africa to Java in just 25,000 years, a mere blink of an evolutionary eye. And two recently redated skulls suggest that these archaic humans reached Java as early as 1.6 million years ago, perhaps even 1.8 million years ago.[36] If subsequent evidence bears out either of these dates, this remarkable migration inevitably raises the thorny question of language.

It is commonly assumed that *H. erectus* possessed insufficient language, and therefore insufficient technology, to have built seagoing rafts suitable for group migration and must therefore have walked to Java during periods of glacially lowered seas—although there is little sign of glaciation severe enough to produce very low sea levels during this early phase of the ice age. The traditionally accepted evidence of language in a prehistoric society is the existence of art and the use of symbols. Until the 1970s the oldest signs of art were in Europe and consisted of cave paintings and some decorated artifacts, none of which were more than 40,000 years old and all of which were exclusively associated with campsites of modern humans. Consequently, most authorities concluded that the development of complex language was confined to modern humans and that it had probably occurred somewhere in Europe between 40,000 and 60,000 years ago.

The Language of Grief

The language debate was given new life in the 1960s, however, when a shallow grave containing the bones of an old man was found in a cave near Shanidar in the Zagros Mountains of northern Iraq. Traces of pollen found in the grave suggest that the man, a heavily boned Neanderthal, had been carefully laid to rest on a bed of flowers and then strewn with more flowers. Analysis of the pollen showed that seven different species had been used to decorate the grave.[37] Such reverential treatment, combined with the symbolism of flowers, suggests that the mourners' belief system included the concept of a separate spiritual existence, and probably an afterlife. According to two separate dating processes, the ceremony appears to have occurred about 60,000 years ago.

In the 1970s, archaeologists uncovered even older evidence of mysticism, this time in a cave near Nazareth in northern Israel. Lying in two shallow graves were three gracile, fully modern human skeletons. One grave contained the remains of a young woman and child, and the other the bones of a teenage boy. The child, who had been about six years old, appeared to have been deliberately interred with the woman, probably to maintain a maternal bond after death. In the other grave, the boy had been laid on his back, and across his chest mourners had placed a single antler from a fallow deer. Bearing neither tooth marks nor other damage, the antler could not have been dragged there later by a hyena or other predator. A small piece of hand-worked ochre was also found in the same sediment layer. Parallel scratch marks on one flattened facet indicate that it had been ground down as a source of pigment either for painting on the cave walls or for body decoration. A group of perforated cockle shells found in the same layer must have been collected from the shores of the Mediterranean, then some 40 kilometers (about 25 miles) away, and were probably strung together to make a necklace or other body ornament. Here was unmistakable evidence that the mourners, incorporating ritual into the burial process, had decorated the bodies of their dead, obviously attaching mystical significance to life and death. Dating by several methods confirmed that the burials had taken place between 80,000 and 100,000 years ago.[38] It represented a serious dilemma for traditional language theorists.

Archaic Australians

All the more intriguing, then, is the discovery of some fifty archaic human skulls, bearing the stamp of a Javan lineage, at widely separated Australian sites.[39] If the message in these Australian skulls is to be taken at face value, then it not only represents an astonishing dispersal for this archaic human but offers eloquent proof of their ability to travel by sea. It also suggests, incidentally, that it was they, and not a descendant form, that first settled the Australian continent. If fully modern human beings had

Skulls: Archaic (Kow Swamp), front; modern (Kielor), rear. The front skull, robust and archaic in style, and the rear skull, gracile and fully modern, were both found in southeastern Australia about 200 kilometers apart, yet both are about 13,000 years old, which means that these two different branches of the human line probably coexisted in Australia for at least 40,000 to 50,000 years. (Casts courtesy of Dr. Alan Thorne, Australian National University, Canberra.)

got there first, these archaic people would not have achieved the wide Australian distribution that their fossils now display. Modern humans would have viewed a subsequent arrival of these archaic forms as an invasion by thoroughly alien creatures and would have promptly set about resolving the problem with the lethal efficiency that is characteristic of our kind. Fierce territoriality is what works best for meat-eating mammals, and having a significant mental edge on their opponents, the original occupants could not have lost. If, on the other hand, the archaics arrived first, they would have been killed off or pushed back by the invading *Homo sapiens,* just as most Aboriginal groups were when confronted by well-armed British colonists tens of thousands of years later.

Certainly the physical differences between the archaics and the later invaders would have been dramatic. The heavy-boned skulls of those first Australians place them well outside the spectrum of modern human beings, and according to anthropologist Alan Thorne of the Australian National University, their skull architecture demonstrates a greater evolutionary distance from modern Aborigines than there is between modern Europeans and the Neanderthals of Ice Age Europe.[40] If there was any cross-fertilization—and judging by modern human behavior there probably was—then it would have been the byproduct of opportunistic rape by raiding parties in most cases, and its genetic impact would have been statistically insignificant. It would have produced little or no change in the morphology or behavior of the conquerors and would show up only in blood groupings and DNA samples. The truly remarkable feature of the archaics' presence in Australia is their durability. The fossil record of those archaic Australians finally peters out in the southeastern corner of the continent only about 9,000 years ago,[41] and represents the last trace of archaic humans, essentially of *Homo erectus* stock, anywhere in the world. Their long residence in this infertile, alien, and inhospitable continent (compared to Java) represents remarkable adaptability by any standards. This achievement alone would have required a relatively high level of intelligence and social cooperation within tribal groups, all the more so in view of the later arrival of competitive modern humans. Add to this their initial migration by sea, and you have an archaic human of astonishing capability.

Sea Barriers

Despite even the lowest sea levels that the most recent ice age has produced, at no time has there been a land bridge between Asia and Australia. In migrating from Java—which formed the southeastern tip of Asia during glacial surges—those archaic explorers would have had to cross several stretches of open sea to reach either New Guinea or northwestern Australia.[42] Traveling to New Guinea via the islands of Sulawesi, Buru, and Ceram, or to Australia via the Wallacean Islands and Timor, even during the lowest sea levels (such as occurred about 140,000 years ago), they would still have faced several sea barriers, some of which were never less than 65 kilometers (40 miles) wide (see figure 16). Even modern boat people, equipped with maps and compasses, would find such a voyage a daunting prospect if they had to entrust their lives to a crude bamboo raft equipped only with paddles to counter the fickle winds and savage tides that characterize this region. That the social cooperation required for such a venture, not to mention the raft-building and seafaring skills, would have been impossible without language clearly indicates that the credit for

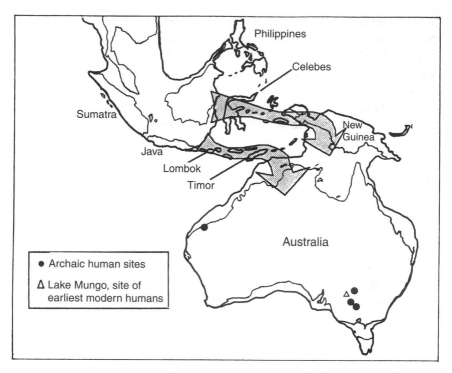

Figure 16. Possible sea routes of early migration to Australia.

the development of language belongs to the *H. erectus* branch of the human family rather than *H. sapiens* and therefore represents an ancient heritage indeed.

Carbon-Pollen Horizons

Exactly when these archaic adventurers reached Australia remains uncertain. The first hint that they arrived very early indeed was unearthed in the sedimentary bed of Lake George, an ephemeral lake in southeastern Australia. In the upper layers of sediment, the percentage of carbon grains—deposited as fallout from local bushfires—is relatively high. Similarly, the pollen grains embedded with them are typical of modern Australia's eucalyptus-dominated, fire-resistant vegetation. Little changes until you dig down to the layers deposited more than 125,000 years ago. In those lower layers, the carbon percentage falls suddenly by 50%, as does the percentage of pollen from fire-resistant plants like the eucalypts. The main pollen types represent older floral assemblies dominated by fire-sensitive species, especially members of the casuarina family (Casuarinaceae) left over from cooler, wetter times. There are clear indications that a warm interglacial period had begun around 125,000 years ago, but this was associated with increased humidity and fails to account for such a dramatic increase in the prevalence of fire in the relatively well watered region around Lake George. The only other comparable jump in the carbon record coincides with European settlement in the area, when farmers began to clear the native vegetation to make way for pasture and cultivation.[43]

Similar carbon-pollen horizons have since shown up in two drill cores taken from the continental shelf east of Townsville and Cairns. The carbon horizon in the northern core sample has been dated to 130,000 years, slightly older than the Lake George sample, whereas an even older carbon-pollen horizon has recently been identified in a drill core pulled from the seabed off the Indonesian island of Lombok. This registered a securely dated carbon peak about 200,000 years ago that also coincides with the beginning of a major change from forest species to grass species in the island's pollen record.[44]

The simple explanation, indeed the only satisfactory explanation, for these well-defined, noncoincidental carbon-pollen horizons and their sequential southward march is that they signal the gradual advance of archaic hunter-gatherers armed with "fire-farming" techniques, techniques eventually copied by fully modern human beings when they arrived in the area tens of thousands of years later. To deny the possibility that those archaic Australians were capable of using fire in this sophisticated fashion would seem to grossly underestimate them. Hearthstones and charred animal bones indicate that fire has been regularly used for warmth, protection, and cooking throughout Africa, Europe, and Asia for at least 400,000 years, and a stone hearth found in a French cave appears to be double that age. Faint evidence even suggests the possession of fire at a campsite near Lake Turkana in East Africa up to 2 million years before that. Moreover, considering the astonishing migration record of those archaic Australians, they fit the intellectual profile admirably.[45]

Another link in the chain of evidence that points to Java as the source of Australia's first settlers was discovered recently on the island of Flores, some 700 kilometers (430 miles) east of Java.[46] Paleoanthropologist Mike Morewood and others from universities in Australia and Indonesia unearthed more than a dozen stone tools in layered sandstone beside an ancient Flores lake bed. Since the tool-bearing strata was sandwiched between two layers of volcanic ash, dating the arrival of humans was a relatively simple process. The two ash layers yielded ages of 800,000 and 880,000 years, so the humans must have occupied the area sometime between the two eruptions. Flores, which is part of the Wallacea chain of volcanic islands, has never been attached to Java, and any migration from Java to Flores would have involved crossing three sea barriers, the largest of which was almost 20 kilometers (12 miles) wide, even during lowest sea levels.

Added together, these factors lead to four inescapable conclusions: (1) the toolmakers were archaic humans of the *Homo erectus* species; (2) they were capable of building efficient watercraft and possessed seafaring skills sufficient for migrating by sea; (3) they therefore must have possessed complex language; and (4) they were actively pushing eastward along the Indonesian island chain more than 800,000 years ago. That they reached Australia within the next 700,000 years is hardly surprising. It would be astonishing if they hadn't.

The real question is, what became of them once they reached Australia? No human fossils or artifacts have been securely dated to more than 60,000 years ago, and there is nothing other than the fifty or so enigmatic skulls and skeletal fragments that can be firmly linked to these archaic humans.

The Body Snatchers

Australia has always been unrewarding for fossil hunters. Most of it is old, flat, arid, and featureless, and very few bones become buried and preserved. But another reason good fossils are relatively scarce is that the continent appears to have supported an unusually large prehistoric population of aggressive scavengers, notably the Tasmanian devil (*Sarcophilus harrisii*) and at least one gigantic lizard (*Megalania prisca*). These two alone would have ensured that remarkably few bodies were left intact long enough to become buried and preserved. The lizard is believed to have grown up to 7 meters (about 23 feet) long and weighed up to 600 kilograms (more than half a ton),[47] about ten times the weight of its modern relative, the giant Komodo dragon (*Varanus komodoensis*), which still roams the islands of Komodo and Lombok where it is feared for its man-eating potential.

The Tasmanian devil, on the other hand, is relatively small and poses no threat to humans, but it is a dedicated scavenger. It eats the entire carcass, fur, flesh, skin, bones, the lot, and was common throughout Australia until the dingo was introduced some 4,000 years ago. Where the devil still roams in Tasmania there are few carcasses that lie around long enough to fossilize. So, the dearth of archaic skeletal material in Australia does not mean that humans were not present in significant numbers long before the ancestors of the Aborigines arrived.

Other clues to human presence such as rock engravings and stone artifacts are notoriously difficult to date, and although the whole continent is decorated with elaborate petroglyphs and littered with knapped stone tools and stone-chip debris left by early hunters, all of this archeological

Komodo dragon (*Varanus komodoensis*). Komodo dragons are primarily scavengers but will attack almost any animal that wanders into close range. Their extinct Australian relative, *Megalania prisca*, weighed ten times as much as the modern Komodo dragon.

evidence is routinely assigned to the Aboriginal culture. No doubt some of this material belonged to the original archaic inhabitants, but until some means is found to distinguish those items from the plethora of more recent material, the enigma is likely to remain.

The oldest firmly dated skeletal material left by modern *Homo sapiens* comes from Lake Mungo in southeastern Australia, suggesting that modern humans first entered the area a little more than 32,000 years ago. But they were invaders, not pioneers. Judging by the Lake George carbon-pollen horizon, the archaics had preceded them by 93,000 years. We also know from their skulls that these archaic humans survived in Australia longer than anywhere else in the world—until about 9,000 years ago at one site in Victoria.[48] What finally sealed their fate we may never know. Perhaps they failed to cope with the postglacial climate change and the desertification of the continent, or perhaps they finally succumbed to attrition by descendants of the Mungo people, the large-brained sharp-witted ancestors of Australia's Aborigines. We can, however, be sure of some

Stone tools found in central Australia. Most of Australia is littered with stone tools and stone chips left by generations of hunters. Unfortunately there is no way of knowing which were left by modern humans (Aborigines and their ancestors) and which belonged to their archaic predecessors, the original inhabitants of Australia.

Shell midden, Lake Mungo, Australia. The remains of campfires and some shellfish meals, a few stone tools, and a 32,000-year-old burial site confirm Lake Mungo as one of the oldest known campsites of modern humans anywhere in Southeast Asia. The lake, now dry, appears to have been abundantly stocked with fish and countless waterbirds at that time.

things: Australia's original inhabitants were adventurous, competent survivors with an efficient technology that included the use of fire. And although they were not modern human beings and left no sign of art or technology that is readily attributable to them, their migratory skills guarantee that they possessed complex language.

The X-Factor

We cannot hope to discover precisely what prompted our *Homo erectus* ancestors to leave their African birthplace and migrate halfway around the world almost 2 million years ago, or what prompted their descendants to cross the sea to Australia. But since we still carry their genes in our bodies it is a fair guess that most of their genetic imperatives live on inside us, imperatives that drive modern human beings on journeys of similar magnitude—to the tops of mountains and the bottom of seas, into the microworlds of molecules, microbes, and nanotechnology, and even out into space in search of new frontiers and our own cosmic origins. The attributes that enable us to do this are of course our unrivaled abilities to communicate, to cooperate, to fashion tools, and to imagine nonexistent events and circumstances. Another, even more vital ingredient is one that we rarely identify, one that plays a crucial role in stimulating the first four and bonding them into a unique and powerful driving force. This fifth factor is our universal tendency to attribute an imaginary specialness to almost everything that comes within the ambit of our perceptions. We are obsessive mysticizers. Focused and magnified by language, this apish streak of imaginative overkill now supercharges almost all human endeavor, fashioning our relationships, feeding our faiths and beliefs, building nations and tearing them apart. It provides the clever neurochemistry that allows us to disengage the rational brain to the point that we can fall in love, pray to our gods, sacrifice our lives, and commit genocide. And it allows us to believe passionately in the patently unbelievable—ghosts, astrology, alien abductions, sustainable development, creationism, and the *X-Files*. The list is endless. It is this genetic X-factor that makes our species at once the most formidable and the most gullible on earth, and it is this mysterious X-factor we will examine in the remaining portion of this book.

Chapter 4

The Agrarian Transition

A Cultural Transformation

Our species' long, slow ascent from the rank of evolutionary fringe dweller to the most powerful and dangerous player ever to stride the biological stage may have been inevitable, but it was never easy. For much of that 2-million-year journey our ancestors' lives were brutish and very short. Even the development of agriculture between 10,000 and 12,000 years ago brought only new hardships, new hazards, and new diseases. However, the behavioral transformation forced on us by the switch to agriculture, industry, and trade, represents not just a watershed in human evolution but, by virtue of its extraordinary consequences, a major milestone in the evolution of life on earth. And in evolutionary terms it occurred only a moment ago.

Until about 12,000 years ago, human beings won their food by methods that might well be described as planned opportunism. At best it was a highly successful strategy, so successful that hunter-gatherers, using age-old hunting techniques, a

Hunters' cave in the Kimberley Plateau, Western Australia. The wall of this concealed cave in the Kimberley region of northwestern Australia is covered with deep grooves worn by generations of Stone Age hunters as they honed their axes in preparation for the next day's kill.

few crude weapons of wood and stone, and a rich fund of accumulated environmental knowledge, survived very well indeed until late in the twentieth century. What they could not do, however, was make the deer give birth to more offspring, the fig tree bear more fruit, or the grasses grow more abundantly.

One solution to this problem was to grow seed crops and edible plants, keep a captive herd of "prey" animals, and build a secure settlement near a reliable water supply. The advantages were immediately apparent: better security from predators and aggressive neighbors, food sources that could be regulated, and surpluses that could be stored. The drawbacks, on the other hand, were long-term and less obvious: the human diet was no longer as rich and varied as it once had been, and the more sedentary lifestyle was not as healthy as the active seminomadic life of hunter-gatherers. Numerous skeletons found in Greece and Turkey indicate a sudden decrease in stature, strength, and health with the spread of agriculture between 10,000 and 6,000 years ago. Analysis of the skeletal material also reveals a significant increase in the incidence of degenerative diseases.

Optimal Health versus Security and Economy

Toward the end of the last glacial episode, between about 14,000 and 11,000 years ago, the average height of the adult human male was 178 centimeters (5 feet 10 inches), and the average height of the adult human female was about 168 centimeters (5 feet 6 inches). By 6,000 years ago, the average height had dropped to 160 centimeters (5 feet 3 inches) for men and 155 centimeters (5 feet 1 inch) for women.[1] This loss of height seems to have coincided with the introduction of farming, and only during the past one or two centuries have we been able to regain the height and health lost at that time. The more sedentary agricultural life did, however, offer a more reliable food supply and greater security from predation, especially for the young.

Agricultural settlement appears to have first occurred near the eastern shores of the Mediterranean, after which it spread fairly slowly through southern Europe and Asia. Where cultural and environmental differences produced variations in technology and regional products, goods were exchanged, which by itself did not represent a revolutionary change in human behavior. Judging by modern hunter-gatherers such as Australia's Aborigines, the tendency for tribal groups to trade goods and technology is as old as our species. Evidence suggests that intertribal barter of tools, ornaments, and ocher was common throughout the Australian continent. But settlement did increase the relative importance of trade. Surplus food and other assets could now be deliberately produced and exchanged for needed items.

Ethologically, however, this relatively minor behavioral shift had ramifications of monumental consequence: the primate *Homo sapiens* was transforming itself into a new kind of animal, an omnivore that no longer gained its food opportunistically but "grazed" at will on homegrown flesh, seed, or foliage; a genetically nomadic species that no longer had to keep moving to survive but could settle in one place and accumulate possessions.

From a Darwinian point of view, this sudden behavioral mutation, unaccompanied by any biological corollary, radically departed from evolutionary precedent. Rather than resulting from a long and laborious interchange between genetic variation and environmental selection, this transformation (misleadingly known as cultural evolution) was brought about by behavior alone. Within the space of just a few thousand years, a new kind of animal had conceived and given birth to itself.

The Transmission of Learning

The ability to learn new patterns of behavior was once thought to be solely the domain of the human species by virtue of its outsized brain and its long period of juvenile brain plasticity. But like all other characteristics considered to be uniquely human, the ability to learn and pass new learning on to others is neither unique to us nor particularly rare in the animal kingdom. Several primates, some sea mammals, and even a few birds are capable of transmitting learned behavior by nongenetic means to succeeding generations. One clear example occurred on a small island off the coast of Japan, where an intelligent young female macaque named Imo learned in 1953 to wash sand from potatoes and other food gifts left on the beach by visiting scientists. Then, in 1954, she learned to separate rice from beach sand by laying handfuls of the mixture on the water, allowing the sand to sink and scooping off the floating rice grains. Other macaques then learned these techniques by observing her. Within a decade most of Imo's tribe had acquired the behavior, and gradually it began to spread to other macaque tribes on neighboring islands.[2] If complex language could have been added to the macaques' talent for observational learning, those behaviors would have spread much more swiftly through the population. In humans, learned behavior disseminates so quickly and passes so easily from generation to generation that the transmission process itself is now revered as one of our species' most valuable and definitive assets.

The Malthusian Mandate

Unfortunately, the human ancestors who developed and benefited from this ability to transfer learning so effectively failed to examine the fine print in the evolutionary contract that accompanied it. That failure allowed the well-concealed penalty clause to remain undetected for almost 10,000 years. In the words of the Reverend Thomas Robert Malthus, an English clergyman and economist writing in 1798:

> I think that I may fairly make two postulata. First, that food is necessary to the existence of man. Secondly, that the passion between the sexes is necessary and will remain nearly in its present state. . . .

Assuming then, my postulata as granted, I say that the power of population is indefinitely greater than the power in the earth to produce subsistence for man.

Population, when unchecked, increases in a geometric ratio, subsistence increases only in an arithmetical ratio. A slight acquaintance with numbers will show the immensity of the first power in comparison to the second. . . .

Through the animal and vegetable kingdoms nature has scattered the seeds of life abroad with the most profuse and liberal hand. She has been comparatively sparing in the room and the nourishment necessary to rear them. The germs of existence contained in this spot of earth with ample food, and ample room to expand in, would fill millions of worlds in the course of a few thousand years. Necessity, that imperious all-pervading law of nature, restrains them within the prescribed bounds. The race of plants and the race of animals shrink under this great restrictive law. And the race of man cannot, by any efforts of reason, escape from it.[3]

Here the danger facing human civilization is spelled out with absolute clarity. Malthus's warning attracted considerable attention, favorable and otherwise, from the church (both Anglican and Catholic), Charles Darwin, and even Karl Marx, and it triggered heated academic and public debate throughout England and Europe.[4]

Malthus was not the first to issue such a warning. Another clergyman, the Reverend Otto Diederich Lütken, rector of a parish on the Danish island of Fyn, published a similar statement exactly forty years earlier:

Since the circumference of the globe is given and does not expand with the increased number of its inhabitants, and as travel to other planets thought to be inhabitable has not yet been invented; since the earth's fertility cannot be extended beyond a given point, and since human nature will presumably remain unchanged, so that a given number will hereafter require the same quantity of the fruits of the earth for their support as now, and as their rations cannot be arbitrarily reduced, it follows that the proposition "that the world's inhabitants will be happier, the greater their number" cannot be maintained, for as soon as the number exceeds that which our planet with all its wealth of land and water can support, they must needs starve one another out, not to mention other necessarily attendant inconveniences.[5]

Despite the controversy initially generated by these statements, the potential threat appeared to lie so far in the future that the argument seemed essentially academic. Submerged by the tide of more pressing political events, the specter of overpopulation gradually slipped from public view until it was finally revisited in the 1960s by more erudite and persuasive writers such as Paul Ehrlich and Julian Huxley. Many might still doubt the immediacy of the problem, but few are now unaware of it.

A central question remains: how could an intelligent species like ours march so blindly into such an ancient and obvious trap?

Plague Mammals

On its face, the problem of overpopulation appears to be a recent one. When agricultural settlement first began 10,000 to 12,000 years ago, probably no more than 4 to 5 million people were scattered round the planet. (Authoritative estimates generally range from 2 to 20 million.) Even when Malthus was writing his first essay on population in 1798, global population had not quite reached 1 billion, having just taken some three centuries to double itself.[6] By contrast, the present population of 6 billion is currently growing by about 80 million a year.[7] Given a quick glance at a graph of population growth over the past 10,000 years—or even the past five centuries—the word *plague* naturally springs to mind.

The world's so-called plague mammals, species that readily achieve exponential (geometric) population growth, include lemmings, rabbits, mice, and rats, and all share four crucial characteristics:

1. They are essentially herbivorous and nonpredatory (although some become omnivorous under certain conditions).
2. They are social animals whose territorial claims are small.
3. They become plague-prone only in unstable environments where food resources and predator numbers can fluctuate dramatically.
4. They are small, highly mobile, promiscuous animals with a very fast reproduction cycle.

Even infertile Australia has a native plague mammal, the long-haired rat, or plague rat (*Rattus villosissimus*), that lives in scattered pockets throughout most of the arid zone. Population explosions of *R. villosissimus* in central Australia are triggered by a drought-flood sequence. A severe drought will occasionally slash the populations of all animal species in a particular region. Should floods then trigger vigorous vegetation growth due to the accumulation of ungerminated seeds, the rats take advantage of the extra food and slip quickly into reproductive overdrive. Meanwhile their main predators, reptiles and raptors, are handicapped by much slower reproduction cycles and cannot keep up with them. Freed suddenly from both starvation and predation, the rat population explodes to plague proportions.

Our species, on the other hand, fails to fit the distinctive four-point profile of a plague mammal on all counts—or so it would seem at first glance.

1. We evolved as omnivores and part-time predator-scavengers.

2. We are tribal and highly territorial.

3. Our food resources are generally stable, and we have no predators to speak of.

4. We are large, semimonogamous animals with a slow reproduction cycle.

But about 10,000 years ago we began altering our ecological profile to such an astonishing degree that by now every one of those characteristics has changed dramatically:

1. By domesticating certain plants and animals and by storing our surplus food, we became essentially a new kind of grazing animal, one that fed not only on vegetable matter but also on high-energy animal protein harvested at will from captive prey. In effect, a single omnivore had secured an option on all of the world's food chains and outmaneuvered all of its natural predators. These primate "mice" now had the keys to the pantry. They lacked only the technology to raid it efficiently. But that would come.

2. By exchanging the rigid social structure of the small, blood-related tribal group for the less regulated society of the agrarian community, our ancestors were in effect swapping the rigid order of primate society for the freewheeling mores of the mouse colony.

3. By settling together in towns and villages and by replacing Stone Age technology with metallurgy, we finally solved most of our predator problems. With the aid of irrigation, fertilizers, agricultural machinery, and better animal husbandry, we succeeded in creating wet-year productivity on an almost continuous basis. But in so doing we released both of the natural safety catches on our fertility potential. The population bomb buried in the human reproductive system was primed and ready for detonation.

4. With the best of intentions the fuse was lit by the giants of modern medicine: Edward Jenner (1749–1823) developed the first vaccine, Louis Pas-

teur (1822–1895) invented pasteurization, and in the twentieth century Alexander Fleming, Howard Florey, and Ernst Chain gave the world penicillin. With the introduction of vaccines, antibiotics, and hygienic medical practices, infant mortality rates plunged, life expectancy increased, and the birth-to-death ratio went haywire.

The Fertility Trap

Unlike rodents, humans did not need a fast reproduction cycle to achieve plague growth. An extraordinarily high fertility potential had already been built into the species during the neotenization process some 2 to 3 million years earlier. By slowing down the species' maturation rate, concealing female ovulation, and extending the life span, evolution eliminated the typically mammalian phenomenon of estrus and expanded the female's window of fertility to more than three decades—an astonishingly long productive period for an animal our size. The social consequences were momentous: Three million years ago a female hominid would have undergone no more than ten periods of menstruation in her thirty reproductive years—just as chimps still do. A modern woman, however, may have a thirty-five-year reproductive window and experience up to four hundred menstrual periods. And since the human female does not come into season or advertise her fertility by behavior, odor, or pigmentation change, she appears to be available for mating from puberty until menopause. Those relatively minor adjustments to our genetic time switches not only produced behavioral changes of extraordinary evolutionary consequence but also pushed the reproductive potential of *Homo sapiens* right out of the chimp category (three to five offspring per female) and into the rodent league (up to thirty per female). With the primate sex drive still running like clockwork in the human male and mortality rates drastically reduced by modern medicine, all the prerequisites for global human plague were in place.

Ancient Witnesses

For most of humanity's 2 million years on this planet, the ratio of fertility to mortality remained fairly balanced, and population growth was relatively slow. Estimates of the global population 12,000 years ago are generally around 5 million or below. By 7,000 years ago, the potential to overpopulate was already inherent, and the population growth rate was climbing steadily. At least two authorities estimate that by 5,000 years ago the global population was about 14 million. The first clear reference to regional overpopulation and its environmental consequences occurs in a poem written on three clay tablets in what is now Iraq a little more than 3,600 years ago. The poet describes how "the people multiplied" until the beleaguered land "bellowed like a bull"; the gods, angered by the uproar, decreed pestilence to reduce human numbers and then flooded the earth with rain (floods being an inevitable consequence of deforestation) to the point that only the hero of the epic survived.[8]

The environmental accuracy of this 1,245-line poem has recently been confirmed during excavations along the Euphrates and Tigris Rivers, in the area once occupied by the Sumerian civilization. Archaeologists have found evidence of the same unhappy sequence of agricultural boom and bust that has accompanied the rise and fall of many cultures. An initial period of colonization, deforestation, agriculture, and irrigation is followed by salinization and agricultural decline, which finally leads to cultural decay and dispersal of the population. Similarly, the Greek philosophers Plato and Aristotle give what appear to be eyewitness accounts of deforestation and soil erosion in two intensively farmed Greek valleys more than 2,300 years ago, accounts confirmed by recent analysis of the local soils and pollen record.[9]

The Black Death

The first sign that the human population was about to launch itself into the exponential growth of the modern plague ironically appeared following the plague that swept through Europe in the middle of the fourteenth century. This bacterial infection, known as the Black Death, was spread by infected fleas living on black rats and came to Europe via the newly opened caravan routes from Mongolia and Manchuria. Europe's population had been rising slowly for several centuries and by the 1340s had reached about 85 million. Between 1346 and 1350 the plague spread quickly to all parts of Europe, and within four years the population had crashed to about 60 million. Yet within 250 years it rebounded to 110 million—approximately the figure it would have reached by that time had there been no plague.[10] Figure 17 clearly illustrates the temporary decline followed swiftly by accelerated growth, a pattern that eloquently expresses the irresistible nature of the primal drive that underpins it—the genetic imperative to reproduce.

Similar rebounds are common in populations of laboratory rats and mice, and the same rule seems to apply to their wild populations. Once a growth curve becomes exponential, the population routinely proceeds to plague proportions, regardless of temporary interruptions. Such evidence suggests that once modern medicine had set the human population jug-

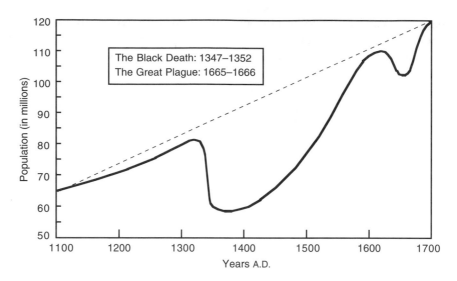

Figure 17. European population growth, A.D. 1100–1700. Modified from Thomas Prentiss, *Scientific American*, 1964.

gernaut rolling, no cultural force on earth could have stopped it. A political regime might try to apply the reproductive brakes by whatever repressive means it chooses, but as soon as the regime changes, the birthrate will again go into high gear, accelerating until it reaches its former trajectory. This rebound phenomenon occurred on a global scale for the first time after World War II, resulting in the baby boom of the 1940s and 1950s. The pattern suggests that once a plague begins, population growth is governed by factors that are largely independent of external influences such as culture or even fluctuating resources.

Growing evidence indicates that many, if not most, mammal plagues do not in fact finally succumb to Malthusian food crises but rather orchestrate their own collapse, just as they regulate their own growth. And in view of the unexpected decline in human fertility during the past fifteen years, it begins to look as though humans are once more proving to be an entirely typical product of the evolutionary process. The human plague has, up to this point, observed all the usual plague rules with astonishing fidelity, so much so that there is little chance our plague trajectory will be altered by the likes of culture and technology.

Our sleight-of-mind transition from hard-pressed, disinherited ape to the most dominant and dangerous species the world has ever seen is clearly one of evolution's most breathtaking achievements. The second stage of that transition, during which *Homo sapiens* auto-mutated from primate hunter-gatherer to plague animal in just 10,000 years merely by applying the novel tools of culture and technology, has no precedent in the natural world. Ethologically, it represents a feat that would have taken genetic evolution tens of millions of years to accomplish. Our overwhelming success, however, has some terrible hidden costs: human activity has now destabilized the biosphere and brought our species to the brink of disaster.

Chapter 5

Evolution's Answer to Biological Waste

Culture versus Evolution

Human language represents much more than complex communication between individual animals. Its development signaled the birth of a major new factor in the spectrum of life's control mechanisms, a force so powerful that it would eventually rival the evolutionary process itself in regulating life on this planet. The fate of hundreds of species may now be decided during lunch. A brief verbal exchange may result in another tract of unique tropical rain forest being reduced to a silent, smoking ruin. It's as easy as that.

Biological evolution and culture seem at first to be a fair match, given our recent success at creating monocultural landscapes inhabited by transgenic plants and animals. Evolutionist Stephen Jay Gould points out that the crucial difference between evolution and culture is in the speed of change. Culture is directed and relatively quick to change, whereas evolution is not; culture tries to construct better systems, whereas evolution proceeds simply by maintaining diversity and weeding out the ineffective. Or, as Gould says, evolution is Darwinian and culture is Lamarckian.[1]

The French naturalist Jean Baptiste Lamarck postulated that organisms were able to acquire physical and behavioral characteristics during their lives, and these characteristics were then translated into genetic information that could be passed directly to their offspring. As it turned out, Lamarck was well off the mark in this case. But cultures do change in precisely that fashion. Societies are constantly modified and redirected by contemporary events, and in many instances they respond to them with breathtaking speed. Europe and Asia are presently characterized by an abundance of rapidly changing societies. The former Soviet Union is one of the most notable examples, its atheistic, monolithic bureaucracy having been replaced in a matter of a decade by political and social discord, widespread unemployment, poverty, crime, and predatory capitalism, accompanied by an upsurge of religion, witchcraft, and other mystic belief systems.

Biological evolution, on the other hand, moves to a very different rhythm. At the molecular level, during the complex processes of DNA replication, it is undirected by the environment and is wholly dedicated to stability. The astonishing accuracy of the replication process itself, in conjunction with its built-in error-correction system, ensures that the spectrum of genetic diversity presented to the environment is produced primarily by Mendelian inheritance and not by transcription error. Out of this narrow spectrum of genetic "hopefuls," only a minute percentage will survive the rigors of the environmental audition before going on to make names for themselves on the evolutionary stage. Evolution proceeds by this inexorably slow and highly selective process, with each species generally acquiring only those features that allow its members to cope more efficiently with the inconstant environment and to launch its genes more effectively into the future.

Biological evolution is not only vastly slower than its cultural counterpart, but it is also regulated by a far more complex and sensitive system of controls, the most crucial of which is the interdependence of species. This interdependence extends throughout the biosphere and represents an organic structure of breathtaking complexity. In some cases the extinction of even a single, seemingly insignificant species may lead to the demise of a whole suite of others that depend on it in one way or another. Pull out two or three such genetic cornerstones and the whole ecological structure may collapse.

The Spreading Plague

None of this was understood when human beings first started to manipulate the local environment to their own advantage. To clear a patch of forest in order to grow food and build a secure refuge must have seemed not only thoroughly harmless but quite beneficial. It became essential for such communities to systematize landownership and land transfer and to regulate the possession and trading of livestock and other material assets. The resulting system, which contributed to peace, productivity, and general welfare, also served to focus and reshape our hominid territoriality and possessiveness into a way of life that would ultimately transform human society entirely. In place of hunting and gathering skills and the traditional wisdom of the elders of the tribe, material possessions gradually became the new measure of social worth and tribal status for most human beings.

In pursuit of this new form of security, status, and power, people were only too willing to divest themselves of their true capital—fertile topsoil, clean air, potable water, and a supportive biota. By exchanging this natural inheritance for short-term cultural advantage, they prospered as never before. And as their material prosperity increased, so did the population. From a world total of fewer than 5 million people 12,000 years ago, it climbed steadily, passing half a billion by 1600, and 1 billion well before Malthus died in 1834. Within the next hundred years it doubled again—in half the time it had previously taken to double itself. Exponential growth was now firmly entrenched and the human plague was under way.

The continual diversions of wars, economic depressions, and technological innovations eclipsed Malthus's warning. Between 1930 and 1960, however, the population increased to 3 billion, and by the mid-1970s it reached 4 billion. *Homo sapiens* was now in full plague mode and growing exponentially at almost 2% a year. The species that had taken 2 million years to reach a population of 1 billion was now producing that number of individuals in less than a generation.

In the 1960s, with population growth still accelerating and per capita food production in serious decline, Malthus's scholarly warning was resurrected by more passionate prophets, notably population biologists Paul Ehrlich and Sir Julian Huxley. The looming crisis they foresaw was averted, albeit temporarily, by the rapid introduction of high-yield crops and intensive farming practices during the 1960s and 1970s. In what came to be known as the Green Revolution, food production leaped ahead of population growth, and most nations, especially those in the West, prospered. Malthus and his modern champions had underestimated the power of human ingenuity and technology—or so it seemed—and their "error" was frequently quoted to illustrate how false it was to equate human beings with other animals.

Even in the late 1970s, the gloomy predictions of "Malthusian pessimists" were still generally discounted, despite the fact that the magnitude and significance of the general biodiversity decline was being accurately assessed for the first time, not to mention the growing menaces of global warming, acid rain, widespread land degradation, and global pollution by fertilizers, pesticides, and other toxic wastes. The extent of the environmental degradation might even then have made the plague connection clear to many more people had it not been for competing appeals for greater productivity that were peddled in all nations under the banners of Growth, Progress, and National Security during those tense Cold War years. Meanwhile, we had ventured into space for the first time and science had begun to unravel the secrets of DNA. The future glittered with promise.

Cultural Blinders

Those who attempt to forecast the future face three heavy handicaps: the brevity of the human lifetime, the anthropocentric bias of the historical record, and the tendency of some biochemical processes to take much more than a human lifetime to respond to culture's assaults. The phenomenon of global warming, for example, brings all of these factors into play. The threshold of perceptible change in a system as massive as the carbon cycle inevitably incorporates considerable prior input, and the symptoms of global warming that began to appear in the second half of the twentieth century were the product of much more than just one hundred years of intensive agricultural and industrial activity. They incorporated the accumulated input of all the land clearing and fossil-fuel burning that humans had ever carried out in the past. Not only did land clearing and fire-farming prime the atmosphere with anthropogenic carbon, but every centimeter of topsoil lost as a result of those activities increased the significance of future soil losses. And as soils become impoverished, plant growth becomes increasingly vulnerable to climate change. In this fashion, the full impact of our present existence will not be completely expressed in the environment for centuries to come.

If an environmental threat takes a human lifetime or longer to clearly show itself, then human culture remains oblivious, continuing on its dangerous course until it plods blindly into the trap. Culture after culture has, in this uncomprehending fashion, fallen victim to the inexorable process of environmental retribution. The Middle East, for example, was not always the semidesert it is today. Although its modern inhabitants are largely unaware of it, they are paying the price for at least 6,000 years of timber cutting and overgrazing by their industrious ancestors. And few if any of those ancestors would have noted the minor incremental changes while they were in progress. Neither would they have understood the causes, nor foreseen the consequences, had they become aware of them.

To Boil a Frog

It is said that if you place an aquatic frog in a pot of water, it will sit there quite contentedly, even if you put the pot on the stove and slowly heat it. According to the story, the frog remains, acclimating for a time to the change in temperature, and just before the water boils, the frog dies. However overstated that claim for frog immobility may be, the analogy holds true for human beings.[2] We lit a fire under the pot 10,000 years ago by clearing the land, sowing seeds, and building communal settlements. The result was inevitable, and retreat is now unimaginable. Like the frog, we are immobile, even though the consequences of our actions—and the pot's temperature—are increasing. We cannot go back. We cling to our current level of technoculture and hope that our ingenuity and technological cunning will somehow protect us from the environmental backlash. Some of us may attempt to lower the heat a little by minimizing our consumption of natural resources and by reducing the volume of waste we produce, but frugality is not characteristic of our kind.

Our waste creates another problem for us. The idea of waste is a purely human construct and a statement of perspective, not fact. In a biospheric and evolutionary context, there is no waste. River water that flows unimpeded into the sea might seem wasted to the irrigator and the hydroengineer, yet every drop of that water has already served a multitude of environmental purposes merely by maintaining the water level and will serve other purposes in the sea and eventually in the atmosphere. The same applies to energy and resources. We bemoan the waste of energy and natural resources that seems to be an unavoidable byproduct of our technology-dependent civilization, but on closer observation we may discover that more efficient production merely generates more activity and greater consumption of energy and resources rather than less waste. In other words, waste is an unavoidable component of biological existence and is even a vital element in the evolutionary process.

False Promises

The personal computer, that ubiquitous modern marvel, came with the promise of a faster, more efficient workplace, even a paper-free office. Instead the result is a whole raft of environmentally expensive associated industries, greater dependence on power supplies, globally expanding communication systems, greater economic complexity, and above all a veritable avalanche of printed matter. Studies of paper consumption in Australian businesses reveal that recent developments in office technology have generated a fourfold increase in the use of paper in the past decade alone.[3] Research in other parts of the world confirms these trends.

Similarly, the invention of better, more fuel-efficient cars simply leads to the production of more cars, larger car factories, more highways, a more mobile population, and increased consumption of mineral resources and fossil fuels to manufacture and run those cars. There may well be minor tactical victories in particular instances of gross negligence, but by and large there appear to have been no strategic environmental gains of any real significance. Every technological breakthrough either generates increased commercial activity or entails hidden penalties that ultimately negate the savings achieved. For example, the catalytic converter, which supposedly eliminates up to 90% of the carbon dioxide, hydrocarbons, and other pollutants emitted by the family car, contains 2 to 3 grams of three rare and precious metals: platinum, palladium, and rhodium. Only Russia, South Africa, and Canada possess commercially viable deposits of these metals. The sophisticated refining processes required to extract these metals from their ores may release up to 10.9 kilograms of sulfur dioxide for each gram of precious metal extracted. The South African and Canadian factories have scrubbers on their smokestacks that remove most of the sulfur dioxide, but the Russian plants lack the expensive scrubbers, and one very large Siberian plant currently emits more sulfur dioxide than all the power plants in the United States combined. When all the greenhouse gases and sulfur dioxide emissions produced by both the factory and the mine are combined, a car containing Russian metals in its catalytic converter has to travel some 25,000 kilometers (15,000 miles), using about a quarter of the working life of the catalyst, before the emissions saved equal the emissions outlayed during the manufacture of the converter.[4] If the car is using the cleaner Canadian metals in its converter, it must then "pay back" the greenhouse gases outlayed to manufacture, fit, and run the scrubbers. Moreover, in the process of eliminating sulfur dioxide and carbon dioxide, catalytic converters generate considerable quantities of nitrous oxide, a greenhouse gas that is some three hundred times more effective than carbon dioxide at trapping heat. Consequently, the rush to install catalytic converters in U.S. cars between 1990 and 1996 produced a 32% increase in the nation's nitrous oxide emissions and is responsible for about half the nation's present overall emissions of the gas. The United States' emissions of carbon dioxide increased by only 9% during the same period.[5]

Oceans of Rust

Waste is not only an integral part of all biological existence, but it also represents a necessary factor in the evolutionary process. Perhaps the most

Hamersley Range in Pilbara, Western Australia. This 2.5-billion-year-old seabed in northwestern Australia is entirely composed of the rusty fallout from iron-rich seas that became oxidized by bacterial wastes. In this sense, the Hamersley Ranges commemorate the greatest pollution crisis of all time.

spectacular evidence of this lies in the Hamersley Range in Western Australia. Representing the geological fallout from the greatest pollution event of all time, these spectacular ironstone ranges are a memorial to life's first major confrontation with the finite resources of a small planet. Viewed from the air, the flat-topped Hamersleys sprawl across the ancient Pilbara Plateau like a cluster of tribal scars, raised and livid. Where erosion has torn open their flanks, their gaudy, iron-rich gravel spills out and pools on the plains like congealed blood. Dusted with green-gold grasses and sprinkled with white-stemmed ghost gums, the Hamersleys flaunt their color with an extravagance that borders on the surreal.

A walk into one of the gorges does little to dispel the feeling of unreality. The smooth pavements and rust-red walls, carved from one of the world's largest banded iron deposits, create the illusion of a deserted alleyway in a somber, rusting city. A small stream slides over the water-polished pavement at your feet and chatters down neat stone steps into a pool on the next level. Here and there, above a green frill of ferns, an old

North Pole fossil found in Western Australia. This tiny column of layered bacterial wastes, representing life's earliest signature on the planet, is part of a 3.5-billion-year-old seabed that now lies exposed on a rocky ridgeline in northwestern Australia at a site whimsically known as North Pole.

fig tree clings with arthritic roots to the iron walls, and birdsongs echo eerily. Touch a magnet to these walls and it will cling to the stone; also notice how the needle of your pocket compass swings aimlessly. You have just stepped through the looking glass of time and now stand at the very heart of an environmental catastrophe that ravaged the planet for half a billion years and entirely changed the course of evolution. The ironstone walls are the remains of a seabed, up to 2.5 kilometers (1.5 miles) thick, that was largely built of bacterially generated rust.

To properly comprehend the magnitude and significance of the evolutionary events whose mark was left here, it is necessary to travel deeper into geological time, almost to the birth of life on earth. If you are already in the Hamersleys, however, such a journey is very easy indeed, since only about 160 kilometers (100 miles) to the northeast, at a site whimsically known as North Pole, you can sit among the world's oldest trace fossils— small piles of debris left on the muddy bed of a tidal lagoon by colonies of photosynthetic bacteria about 3.5 billion years ago.

Two Streams of Life

Biological evolution had been operating for more than half a billion years by the time these tiny bacterial signatures began to appear along the seashores of the world. We know relatively little of the life of that period except that a single prototypic organism had successfully given birth to two separate streams of life, one consisting of eubacteria and the other a genetically distinct group of bacterial organisms known as archaebacteria, or archaea.

Eubacteria lived wherever there was water, extracting chemical nutrients from their surroundings, in some cases harnessing the sun's energy to synthesize the nutrients they required to grow and reproduce. To do this they needed a highly reactive fuel, and the natural choice in that oxygen-free environment was hydrogen. Most early photosynthesizers appear to have mined their atomic hydrogen from hydrogen sulfide gas, using sunlight to break its molecular bonds. A little less than 3 billion years ago, however, some of those photosynthesizers upgraded their photochemistry to one that enabled them to extract their hydrogen fuel from an even more abundant raw material—water. This new strategy presented them with a far more efficient method of powering their life processes, but like all innovations, high yield also meant high cost.

The attached fee was small, but ominous. Although the splitting of each water molecule resulted in two atoms of valuable hydrogen, it also left one atom of highly reactive oxygen. So long as the density of bacteria remained low, the volume of waste was negligible, but by 2.8 billion years ago the numbers of these oxygen producers had increased dramatically around the world—with predictable consequences.

During the next half billion years, the eubacteria's oxygen wastes not only polluted all the world's oceans but also began to leak into the atmosphere. At the same time, earth's ancient seas, which had been loaded with soluble iron, turned red and cloudy with nonsoluble iron oxide—rust. Unbridled evolutionary success, and the massive waste that it invariably entails, had polluted most of the biosphere. Life would never be the same again.

By the time the rust settled to the seafloor and the waters finally cleared, a new spectrum of organisms had evolved. Some had evolved ways of avoiding or protecting themselves from oxygen, and some had learned to live with it and even use it. Among the latter was a group of archaeal-eubacterial composites that coped more easily with the growing problems of starvation and oxygen pollution. What most distinguished these composite cells, known as eukaryotes, was their peculiar internal structure. The cell's DNA was sequestered within a finely perforated double membrane envelope, and several bacterial lodgers now lived symbiotically within the cell. These bacteria, originally predatory or parasitic, appear to have called a truce with their host and set themselves up as a collaborative consortium of service providers in return for refuge from the hazardous environment outside.[6]

Our Oxygen Police

Two billion years later those symbiotic bacteria are still there, serving in their various roles, in every cell of our bodies. One group, known as mitochondria, now serves as our cells' oxygen police and fuel suppliers, and it is because of them that we are able to breathe and to move. They gobble up any oxygen that penetrates the cell's outer membrane and use it as a high-octane additive in the adenosine triphosphate fuel they manufacture.

It could be argued, therefore, that mitochondrially powered, oxygen-breathing animals like us are living memorials to that original pollution event. Its scars remain plainly visible, at both the cellular and global level, showing up under the microscope as the complex internal structure of the nucleated cell, and globally as the third branch on the tree of life, the eukaryotes. Two kingdoms of eukaryotes would eventually colonize most of the planet's crust and now appear to dominate it, thanks largely to their peculiar ability to form huge, highly specialized corporate groups that live and die as one. Those two kingdoms of corporate eukaryotes are, of course, plants and animals. In other words, even we, as bacterial–archaeal composites, bear the marks of that original bacterial plague.

It is worth reminding ourselves at this point that the apparent domination of the planetary surface by plants and animals is thoroughly misleading in every respect. Not only do those two kingdoms represent just two small twigs on the massive tree of life, but they also contribute less biomass to the global biota than do microbes and exert far less influence on the biosphere's chemical cycles.[7]

The Turning Point

There has never been another pollution crisis to match the oxygenation of the biosphere. Originally absent from the earth's atmosphere, oxygen now makes up more than one-fifth of it. This accumulation of bacterial oxygen in the biosphere would have been nothing short of catastrophic to the life forms that preceded it, yet without it we would not exist. It was the leakage of bacterial oxygen from the sea into the atmosphere that turned earth's skies blue and built the radiation shield that eventually allowed larger life forms to venture from their watery birthplace and colonize the high ground. And it is that same bacterial oxygen that now fuels the muscular machinery of the entire animal world.

In this fashion, then, the toxic wastes of one biological era provide the springboard for the next. Even human waste may be destined to have a similar effect, on the one hand undermining the systems on which much of life now depends, and on the other, launching evolution in a multitude of new directions.

Another strange analogy between the explosive proliferation of bacteria some 2.8 billion years ago and the recent explosive growth of the human population is that each of these evolutionary events was triggered by a minor modification to a preexisting structure, a modification that added nothing tangible to the morphology of the organism, such as an extra flagellum for the bacteria or a better set of fighting teeth for our hominid ancestors. Yet in both cases it was as if evolution had switched on a supercharger. For the bacteria, this supercharger lay within its photosynthetic chemistry; for humans it lay in the language areas of the brain.

The Price of Success

Let's consider the bacterial scenario first. The seeds of that catastrophic success were sown on the warm, muddy fringes of primordial seas and lakes all around the world more than 3.5 billion years ago, when some sulfur-eating bacteria developed ways to use sunlight to extract hydrogen from the mineralized waters in which they lived. It was an economical and efficient means of subsistence, although inevitably it left a small quantity of wastes. Getting rid of waste gas was easy, but where photosynthetic bacteria lived together in large groups, the wastes accumulated beneath them to form layered mounds known as stromatolites.

This is the chemical origin of those earliest trace fossils that now appear on the ridgeline at North Pole in Western Australia. In all probability those early stromatolites also represent the product of a precursive form of chlorophyll molecule. The best evidence for this lies about 800 kilometers (500 miles) to the southwest, in the warm, saline recesses of Shark Bay. There, modern photosynthetic bacteria build precisely the same kind of structures. The Shark Bay stromatolites are up to a meter (3 feet 3 inches) across and in some cases grow more than a meter tall.

Despite their orderly appearance, stromatolites are in fact simply high-rise bacterial rubbish dumps. Each day the bacteria that colonize the upper surface of the stromatolite extract mineral nutrients from the water with the aid of sunlight. Then they dump their calcium carbonate wastes, extricate themselves from the sticky, silt-laden deposit, and migrate to the surface once more. In this fashion, year by year, millimeter by millimeter,

Underwater stromatolites found in Shark Bay, Western Australia. Precisely echoing the structure and origin of Western Australia's North Pole fossils, these stromatolites, still being added to by colonies of modern cyanobacteria, line the shallows of Shark Bay.

their stromatolitic waste piles grow. Those at Shark Bay represent by far the oldest and largest population of "live" stromatolites anywhere in the world.[8]

The first signs of the metabolic upgrade that ultimately polluted the whole planet with oxygen—and still expands the Shark Bay stromatolites—can be seen barely 100 kilometers (60 miles) from the North Pole fossils. Should you walk up a certain small ravine that winds into the scrubby, undulating country to the north of the Nullagine River, you may suddenly become aware that there is something very odd about the dark cliffs to your right. They consist entirely of densely packed, sectioned domes, some up to two meters in diameter, astonishingly symmetrical, and all of them finely layered like massive black onions. These, too, are stromatolites, and they are about 2.8 billion years old.

In rocks older than this, stromatolites are relatively rare and invariably small, but beginning 2.8 billion years ago they become massive and very common in the fossil record. The general consensus among paleontologists is that this marks the point at which bacteria put the finishing

Nullagine fossil found in Pilbara, Western Australia. Signaling the explosive success of a water-fueled chemistry based on photosynthesis, this giant stromatolite in north-western Australia is about 2.8 billion years old. The chemistry that built this stroma-tolite eventually yielded sufficient oxygen to rust the seas, turn the earth's skies blue, and permit the development of oxygen-breathing animals like us.

touches to their complex photosynthetic processes and in return reaped a disproportionate reward. With the aid of a minor upgrade of their chem-istry, these photosynthetic bacteria were able to multiply their original energy budget by a factor of 18. It was also to be their undoing. The inno-vation would not only launch them into global dominance for almost half a billion years, but it would also plunge them into the most catastrophic plague the world has ever known.

The Human Analogy

The explosive rise of human beings during the past 10,000 years presents an ominous analogy. Our genetic lineage seems to have barely survived for the first half million years of its existence. During the next 2 million years, however, the evolutionary tide gradually turned. By 1.6 million years ago, our ancestors had spread halfway around the world, and by

Darling Harbour, Sydney, Australia. Whatever human beings leave behind as a by-product of their existence is their natural biological waste, just as stromatolites, iron deposits, and oxygen-blue skies are the biological waste products of photosynthetic bacteria.

New York, New York. This prodigious accumulation of biological deposits on the banks of the Hudson River appears to echo the monolithic uniformity of the Shark Bay stromatolites, yet it incorporates an unprecedented variety of waste materials.

11,000 years ago, with the colonization of the Americas, they finally achieved a semiglobal distribution. About that time, too, hunter-gatherers were beginning to experiment with agriculture and to form farm-based settlements. Within the next few millennia towns and cities would spring up on all continents; culture, commerce, and technology would become primary preoccupations; and the species would launch itself into plague, eukaryote style.

Yet despite this dramatic behavioral transformation, *Homo sapiens* had undergone no morphological change whatever for at least 100,000 years, and only very minor changes in the past 2 million. So since the primary behavioral distinction between ourselves and our closest genetic relative, the chimpanzee, is the volume and complexity of our verbal communication, there can be little doubt that this was the catalytic factor most responsible for our explosive success. Like the development of oxygenic photosynthesis in bacteria, the development of complex language enabled us to dramatically increase our resource consumption. Consequently, one of the distinguishing features of our plague is the volume and diversity of our

High-rise car cemetery, Queensland, Australia. All biological activity, including ours, results in various levels and forms of waste, not all of which is considered toxic.

biological wastes. Whereas cyanobacteria left behind a few limestone reefs, massive iron deposits, and earth's oxygen-blue skies, our wastes include skyscrapers, shopping malls, motor vehicles, freeways, huge volumes of greenhouse gases, oxides of nitrogen and sulfur, poisonous metallic compounds, radio-active wastes, DDT, organochlorines, and a vast array of artificial substances both toxic and nontoxic. That some of our biological wastes are poisonous to most forms of life is also entirely typical of earthly existence. The oxygen producers of 2.5 billion years ago ran into exactly the same problem. In the words of evolutionary biologist Lynn Margulis and Dorion Sagan, "There is no life without waste, exudate, pollution. In the prodigality of its spreading, life inevitably threatens itself with potentially fatal messes that prompt further evolution."[9]

Blinded by our anthropocentric view of the world and armed with the technology to wreak astonishing damage on a global scale, we are fulfilling all of the normal evolutionary requirements of a plague species, and the environment is responding precisely as it always has—from the polluter's viewpoint, of course, it is responding badly.

When photosynthetic bacteria were rare in the world's oceans, their toxic wastes were of little consequence. As their plague climbed slowly toward its peak around 2.2 billion years ago, the biosphere rolled out its juggernaut of environmental retribution and began to re-level the evolutionary playing field. Today, as environmental degradation mounts and biological diversity plummets, there is every reason to suspect that the ancient juggernaut is rolling once again.

Chapter 6

Correcting Imbalances

Life is trouble.
—Nikos Kazantzakis, *Zorba the Greek*

Earth's Biological Skin

The earth is, in effect, a cosmic organism with a biota and biosphere that continually interact with the body of the planet and with the cosmos as a whole to maintain viable conditions for the life that exists here, argues British atmospheric chemist James Lovelock. The proposition, as stated this way, is Lovelock's Gaia hypothesis, named for the ancient Greek goddess of the earth. Lovelock suggests that it is no longer sufficient to propose, as Charles Darwin did, that "organisms better adapted than others are more likely to leave offspring." The Gaia theory adds the rider that all organisms, by their chemistry, behavior, and interaction, tend to modify the environment to increase the chances of their continued existence. It proposes that the evolution of all species, and the evolution of the biosphere they inhabit, is thereby tightly coupled and operates as a single mechanism.[1]

In many ways this theory echoes a proposition advanced earlier in the twentieth century by Russian geochemist Vladimir Vernadsky, in which he described all elements of the biosphere as "living matter" that constantly interacts to form a single gigantic ecosystem. It has even been argued that, in view of the close relationship and constant interaction between the biosphere and the body of the planet, it is the biosphere, and not the crustal material, that represents the outer skin of the planet. In the words of biologist Lynn Margulis and coauthor Dorion Sagan, "Taken at its greatest physiological extent, life *is* the planetary surface. Earth is no more a planet-sized chunk of rock inhabited with life than your body is a skeleton infested with cells."[2]

Conventional wisdom has it that if it were not for earth's two main greenhouse gases, carbon dioxide and methane, the mean temperature of the planet at sea level would be some 33°C lower than it is. And consequently, if the earth were suddenly to shed its biologically maintained blanket of atmospheric carbon, all life would be extinguished and the planet would swiftly return to its original state—a barren sphere of icy rock. But, as Lovelock points out, these images are misleading and the relationship between the planetary mass and the biosphere that envelops it is much closer than casual observation might indicate.

It is true that the primary moderator of earth's temperature is carbon and that carbon is the linchpin of the chemistry of life and is continually recycled by earth's biota. Carbon is present in the atmosphere primarily in three forms: carbon dioxide, methane, and—a recent addition—synthetic chlorofluorocarbons, or CFCs. Together with water vapor, these are the principle greenhouse gases. It is also true that these gases absorb some 88% of the infrared (heat) energy radiating from the earth's surface, thereby significantly warming the thin shell of atmosphere that surrounds us. Lovelock points out that just before life appeared on earth, the atmosphere contained vastly more than its present level of carbon dioxide (currently about 0.035%), and the global temperature was far higher than it is today. Earth's two lifeless neighbors, Mars and Venus, have an oxygen-free atmosphere that is 90% carbon dioxide, whereas our present atmosphere is 20.9% oxygen and only 0.03% carbon dioxide.[3] It is generally agreed that the source of that difference is bacteria. It was earth's primordial bacteria that originally removed the carbon and added the oxygen, cooling the planet and making it generally habitable. That pleasant state of affairs is still maintained by the world's bacteria, even those now living as plastids inside other organisms. For example, the photosynthetic cells of all green plants are bacterial plastids, as are the mitochondria in our cells.

This Cosmic Camelot

Earth's modern photosynthetic organisms therefore constitute not a heating mechanism but a gigantic air conditioner. They cool the planet by

Napier Range, northwestern Australia. Earth's most spectacular long-term reservoirs of biologically sequestered carbon are its reef systems. In combination with calcium and oxygen, billions of tons of carbon have been locked away in this fossilized reef system in northwestern Australia for about 360 million years. Known as the Napier Range, this ancient barrier reef once fringed the entire Kimberley region, making it almost as large as the Great Barrier Reef off the coast of northeastern Australia.

pumping carbon from the atmosphere into their bodies and from there into the soil, rocks, and oceans, where it may remain locked out of the carbon cycle for long periods of geological time. In this fashion, these primary producers have kept the blanket of atmospheric carbon dioxide thin enough to allow some 12% of the earth's radiant heat to escape, thereby maintaining an optimum environment for the biota as a whole and turning the planet into the cosmic Camelot we have inherited.

Because of its extreme sensitivity to environmental and biological changes, this carbon pump provides an excellent thermostat for the whole system (see figure 18). High levels of atmospheric carbon dioxide induce greenhouse warmth and humidity; the carbon-extracting multiplication of photosynthetic bacterial organisms (such as cyanobacteria and the chloroplasts in leaves) promotes a flourish of predatory organisms, and consequently of all life. This accelerating cycle of life pumps carbon dioxide out of the atmosphere, into the biota, and then into the earth's crust, and cooling occurs. Ice

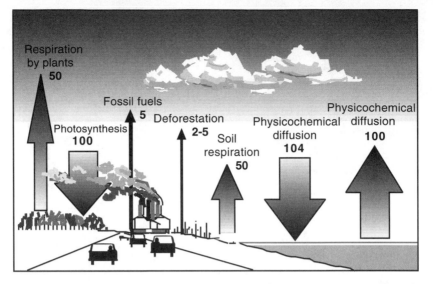

Figure 18. The carbon cycle. Annual carbon fluctuations are shown in units of one billion metric tons. The net atmospheric gain is between 3 and 6 billion tons annually. (From multiple sources.)

ages are dry times, characterized by low levels of atmospheric carbon dioxide and water vapor. Photosynthetic bacterial activity is consequently inhibited, thereby reducing carbon extraction, and carbon dioxide accumulates in the atmosphere. More heat is then retained and the system rebalances itself.

The amount of carbon an organism extracts from the air during photosynthesis and the amount it returns during the processes of respiration and its final decay are roughly equal. However, the amount of carbon exchanged on an annual basis is so large that even the smallest percentage of imbalance, in one direction or the other, results in a significant change in the flow of carbon in or out of the atmosphere. Land plants, for example, recycle about 100 billion metric tons of atmospheric carbon each year, only about 14% of the current atmosphere's total carbon content. But by virtue of plants' sensitivity to environmental change, this 14% fluctuates quite a lot and represents a powerful control factor in the global environment. A temperature change of only 1°C can alter a plant's respiration rate by up to 30%.[4] Such highly sensitive control factors offer evolution—and human beings—very large levers to play with.

The Methane Lever

Methane, however, is one major bacterial component in the carbon cycle that lies well beyond human control. Isotopic analysis of methane from various parts of the biosphere suggests that most of it comes from natural sources, the main ones being archaea, or archaebacteria, the little-known second branch of earth's original tree of life. Archaeal bacteria still infest the planet, but unlike their eubacterial counterparts, they are now very well hidden. What they are hiding from is oxygen. Having evolved long before earth's biosphere contained any free oxygen, most archaea, especially those that fed on hydrogen and produced methane waste, had never devised any chemical means of coping with oxygen's highly reactive atoms. Many eubacteria already used metabolic processes that yielded oxygen as a byproduct, having learned early how to cope with, and even incorporate, oxygen into their metabolic processes. Consequently, when eubacteria flourished and began to charge the seas with their highly reactive oxygen wastes, oxygen-sensitive methane producers were faced with only two alternatives: take cover or perish.

Perhaps reflecting an origin in underwater volcanic vents or in volcanically heated lakes and mud pools, many methane producers, or methanogens, could already cope very well with environmental extremes. Some, known as hyperthermophiles, come into their own at temperatures that tear other biochemistries apart. In fact, the record for hyperthermophilia is currently held by an archaeal organism known as *Pyrolobus fumaris*, which has been found to thrive at 113°C, well above the boiling point of water. Two billion years ago such talents would have enabled methanogens to retreat to habitats where even oxygen could not follow. Some remained in hot springs and submarine volcanic vents, but most went underground, either literally or figuratively. This is one reason scientists only discovered them relatively recently. Once they had been discovered, however, researchers began to recognize their peculiar chemical and genetic fingerprints everywhere, even in the intestines of methane producers such as cattle and termites. But the real surprise came during biological analysis of drill cores, which showed that vast numbers of methanogens also occupy interstitial habitats in fracture lines and sedimentary rocks at astonishing depths. According to University of California researcher Norman Pace, increasing evidence indicates that "the crust of the Earth is shot through with bacterial biomass, wherever the physical conditions permit."[5]

Bacteria have even been found embedded in rock 3.5 kilometers (almost 2.2 miles) down in a South African gold mine, and some 10,000 strains of organisms have been recognized in drill cores from the southeastern United States. Metabolism of hydrogen is the dominant theme in such places, and methane is the predominant product. The metabolic rates of these microorganisms are extremely slow, however, and they may reproduce only once every few hundred years, but the sheer volume of bacterial life means that their methane output is prodigious. Some researchers

Glacial dropstone, central Australia. Three separate layers of glacial debris scattered across Australia's desert heartland provides graphic evidence of the unstable nature of earth's climatic patterns. This ice-polished boulder fell from the underside of a marine ice shelf some 750 million years ago when the continent lay beneath a sheet of ice as thick as the one that covers Antarctica today.

even argue that the world's gas and oil reserves are wholly the product of these organisms. Indeed, isotopic analysis shows that most of the methane in the outer few kilometers of the earth's crust and at the surface has been produced by bacteria. The underground density of methanogens has led some researchers to believe that their global biomass not only rivals that of all other bacteria but may even exceed that of all other earthly life.[6]

Defrosting the Freezer

Of greatest significance to us, however, are the vast populations of cold-loving methanogens that inhabit the polar and subpolar underworld. Bacterially produced methane, imprisoned in permafrost and icy marine hydrates, as well as the methane that would be produced by the anoxic decay of vegetation locked in tundra permafrost, represents a carbon time bomb of awesome potential should global warming unlock the earth's iceboxes. If this sounds like scaremongering, we should remind ourselves that human

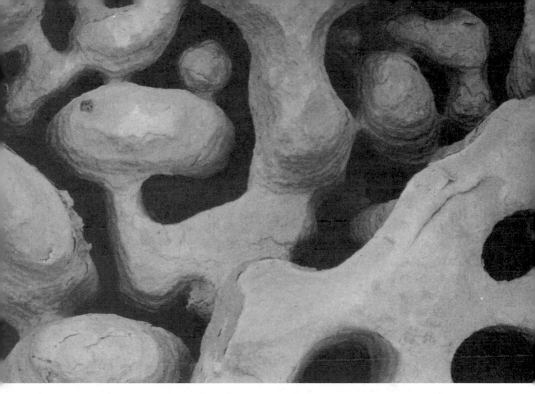

Sandstone formations in northwestern Australia. Winds eddying within a rock shelter have worn down the softer parts of this sandstone layer to reveal the original turbidity patterns in the glacial meltwater that deposited the sediments some 280 million years ago.

records go back only a few thousand years—in geological terms, the blink of an eye. In actuality earth's climate has tended to fluctuate quite a lot. Two layers of glacial debris scattered across Australia's desert heart bear witness to dramatic climate shifts in the past and belie historical records of climatic stability. (Australia lay in tropical latitudes during both glacial periods.)[7]

There now seems to be little doubt that the temperature reversals that characterize ice ages often occur with astonishing abruptness—almost as though the global climate were governed by a binary switch. Recent research indicates that some of the transitions between glacial and interglacial climates may have occurred in less than one hundred years, and perhaps only a decade or so in some cases—almost fast enough, as one meteorologist put it, to capture the attention of politicians. Although there is a close statistical correlation between carbon dioxide levels and temperature throughout the glacial record, "the temperature changes are from 5 to 14 times greater than would be expected on the basis of the radiative properties of carbon dioxide alone," according to Richard Houghton and George Woodwell, scientists at the Woods Hole Research Center in Massachusetts.[8] And since, molecule for molecule, methane is many times more effective as a greenhouse gas than carbon dioxide, there seems little doubt that the release of large quantities of methane into the atmosphere could have devastating and far-reaching effects, powerful enough to produce one of these dramatic reversals in climate. We awaken the sleeping giant of tundra methane at our peril.

Apart from its central function as the atomic pivot on which all organic chemistry turns, carbon plays two other vital roles. Combined with oxygen as carbon dioxide it provides the principal food source for all photosynthetic organisms, thereby underpinning 99.9% of all life on the earth's surface; it also provides the primary thermostat by which life regulates its own environment and contributes to its own survival.

A Self-Governing Planet

The self-regulatory carbon cycle is, according to Lovelock, symptomatic of all of life's interactions with the planet. It was this that led him to conclude that the planet represents the cosmic equivalent of an organism, maintaining and regulating its biospheric environment to produce optimum conditions for life's survival. The *prima facie* case Lovelock established for the existence of this overarching mechanism also offers a plausible explanation for the nature and global scale of the environmental disturbance that has by now become apparent.

In testing his original proposition, Lovelock constructed a computer model of a world inhabited only by two species of daisy, one dark and one light, each with slightly different effects on the environment due to its different reflectance value. Given that this world was earth's twin except for its lack of biodiversity, and provided that earth's normal feedback mechanisms were allowed into the equations, the daisies invariably survived very well. They competently managed the environment to their own advantage, stabilizing their Daisyworld environment at the optimum level for their joint survival.

Lovelock's unavoidable conclusion was that the primary characteristic of this hypothetical biosphere was its stability. Neither of the daisy species was able to overwhelm the other. If one did begin to reproduce more rapidly than the other, the environment automatically became unfavorable to it and its population growth was curtailed. Lovelock then gradually added other species, including daisy-grazers and their predators, and subjected all to occasional environmental catastrophes such as plagues, ice ages, and comet impacts. In some instances up to 30% of the daisies were destroyed. In every case, however, all species recovered quickly and the temperature restabilized—even when Lovelock factored in a gradual rise in the sun's temperature.[9]

Apart from a certain discrepancy of scale, evidence suggests that similar mechanisms are at work in the real world. As part of the regulatory processes that maintain earth's temperature and biodiversity, a severe penalty is imposed on any species that destabilizes its immediate surroundings, and as a result disrupts the hard-won balance between the earth's biosphere and its biota.

Paying Our Debts

Humans have done exactly that; we have destabilized the environment and disrupted the balance between the biosphere and its inhabitants. We have dallied too long at the banquet of natural resources, only to discover that the only way out is past the cashier. Even among those who are aware of the scale of our environmental debt, the general consensus seems to be that with the aid of a little fast technological tap dancing most of us might yet make our escape without paying the full price. Not only would this involve a drastic and immediate reduction in the daily rate at which we gobble up the world's energy resources and dump our wastes, but we would have to sacrifice two of Western civilization's most sacred cows—Growth and Progress—to do it. Sadly, nothing in our history suggests that we would be able to accomplish such a feat—at least not until it is far too late to have any effect on the outcome.

Even in sparsely populated Australia, where land degradation is already visible in varying degrees over most of the continent and the rate of mammal extinction is among the highest in the world, the talk is still of population growth and sustainable development—concepts as rational as perpetual growth in a finite system. Yet such myths are still graphically enshrined in textbook flowcharts that make no reference to the environmental costs of human activity, especially economic activity. By perpetuating these omissions, economists are even able to view major oil spills and other pollution catastrophes as industrial opportunities and therefore claim their cleanup costs as part of the gross national product.

Theologically, of course, the economic rationalist, the industrialist, and the developer all have a watertight defense:

So God created man in his own image. . . . And God said unto them, "Be fruitful, and multiply, and replenish the Earth, and subdue it: and have dominion over the fish of the sea, and over the fowl of the air, and over every living thing that moveth upon the Earth."

—Genesis 1:27–28

Mine pollution, Queenstown, Tasmania. These barren mountains around Queenstown in western Tasmania were clothed in dense temperate rain forest until mining began in the 1880s. The modern devastation is the combined product of many decades of logging and toxic mine fumes. It represents Australia's worst environmental disaster.

> And he hath constrained the night and the day and the sun and the moon
> to be of service to you, and the stars are made subservient by His command. Lo! herein indeed are portents for people who have sense.
>
> —Koran, Surah XVI

Such divine commandments have always represented sound logic in the mystically fertile marketplaces of the commercial world, and it is hardly surprising that Charles Darwin and Alfred Russell Wallace received such bad press when they tried to warn their contemporaries that they were descended not from heaven but from an apish ancestor. You did not have to be a religious extremist to take offense at that. Darwin and Wallace were propounding not only a spiritual heresy but also an economic one. So long as we believed that our origins were divine and that we therefore owned the real estate, we remained free to deforest, farm, mine, and pollute at will, without even contemplating, let alone budgeting for, the ultimate environmental cost of these activities.

True Believers

It was inevitable that the creation-evolution argument would be long and bitter. Even today in the United States, pollsters find that more than half of their respondents believe in the existence of the Devil, and only about 45% of the population is willing to concede that their origins might be other than divine. Four hundred years after Copernicus, and despite the considerable achievements of the National Aeronautics and Space Administration, 54% of Americans still believe that the sun revolves around the earth.[10]

Although Darwin's theory of evolution and natural selection embodied the most scientifically productive insights ever achieved, it also constituted one of the most heretical and revolutionary propositions ever advanced. Consequently, as a doyen of British science, Darwin inevitably drew far more fire than Wallace, suffering insult and derision from such eminent authorities as Bishop Samuel Wilberforce, pillar of the Anglican church and staunch defender of orthodoxy; Adam Sedgwick, professor of geology at Oxford; and Britain's most famous anatomist, Sir Richard Owen.

Darwin had his champions, of course, but they were a very mixed bunch. As well as the redoubtable T. H. Huxley, the list soon included Karl Marx, Friedrich Engels, and even one Adolf Hitler. Marx and Engels enthusiastically embraced the concept of "survival of the fittest" as the philosophical underpinning of their political platform. Hitler, too, seized on a similarly distorted version of "Darwinism" as the philosophical foundation for Nazism, and regurgitating it in the guise of eugenics, used it to justify one of the worst genocides the world has ever known.

As the twentieth century unfolded, the evidence in favor of evolution mounted, and the genetic origins of our species became harder to deny. Nevertheless, the corollary of Darwin's evolutionary theory, that we are not owners of the real estate, and not even joint tenants but tenants-in-common (i.e., all species together own the whole of the real estate), remained generally beyond our grasp. Consequently, aided by technology and a burgeoning population, humanity continued to squander its natural capital with greater ardor and efficiency than ever before. With one species in plague and most of the others in decline, this Daisyworld was now thoroughly out of balance.

The Bitter Pill

The remedy for such imbalance is as simple and effective as it is inevitable: the plague brings about its own collapse, the biota rebuilds itself, and life goes on. If we were truly rational creatures, that is to say, thoroughly atypical and even unnatural by earthly standards, we might have been able to beat a hasty retreat. But that is not how our DNA plays out. According to Lovelock's Daisyworld model, the only alternative is unpleasant indeed: one way or another some of us will have to go. And as far as the evolutionary health of the planet is concerned, the sooner this occurs the better. In fact, stabilization at present population levels may very well represent a worst-case scenario. Fortunately (from a biospheric point of view), we are already doing everything possible to avoid that dangerous possibility. For example, in the wealthier nations of the world, where humanity's rush to the population precipice now shows significant signs of easing, we attack the death rate and boost the birthrate with every medical weapon in our considerable arsenal. Voluntary euthanasia and abortion are widely prohibited, and human life is preserved at any cost.

For those couples who are physically unable to contribute to the population catastrophe in the traditional fashion, our technology offers the possibility of medical miracles in the form of the sperm bank, the frozen embryo, in-vitro fertilization, and even cloning. Although in vitro procedures might currently yield an inconsequential number of additional births and represent an insignificant addition to humanity's total impact on the global environment, they clearly betray the raw, irrational power of the fundamental instinct that fuels the human plague: the genetic imperative to reproduce.

The Chemistry of Collapse

Many other animals seem to have a hormone-regulated response to environmental stress that switches their metabolism into a more economical mode whenever resources become scarce. Inevitably the energy-hungry processes of reproduction are the first to be targeted. For example, Australia's kangaroos are able to put the development of their embryos on indefinite hold during drought by means of a unique, hormone-regulated phenomenon known as embryonic diapause. When the drought ends and food becomes readily available once more, the embryo resumes normal development. Wild populations of yellow baboons (*Papio cynocephalus*) have also shown signs of hormone-regulated fertility decline during periods of social and environmental stress. The telltale hormonal signature of this process, first detected by endocrinologist Sam Wasser of the University of Washington, has also been identified in captive lowland gorillas, and in women. When persistent in women, it typically accounts for up to 10% of infertility and 25% of habitual abortion.[11]

A similar hormonal signature with even more dramatic outcomes has been recognized in other species, notably snowshoe hares, lemmings, voles, rats, mice, and tree shrews. These animals, however, are more likely to become infertile during periods of social rather than environmental stress, and the primary governing factor seems to be population density. When the number of individuals rises to a certain level, a suite of internal factors, all hormone-regulated, appears to cut in, limiting fertility even where food resources remain unrestricted. This process has been shown to lead to a total cessation of reproduction in some cases, and occasionally leads to population extinction, although how this occurs is still poorly understood. The most startling feature of these collapses is that even when numbers had plummeted to the point that density was low and social stress was no longer a problem, instead of fertility rising and death rates falling to allow the population to recover, the decline accelerated.[12]

General Adaptive Syndrome

This curious mechanism, originally called the general adaptive syndrome by Hans Selye, who first noted it in 1936, has since been reported in several wild rodent species, but it is best documented in laboratory rats.[13] In well-fed but extremely dense populations, social stress appears to trigger a broad spectrum of abnormal physiological responses such as glandular malfunction, inhibited sexual maturation, diminished ovulation and implantation, inadequate lactation, increased susceptibility to disease, and a sharp rise in infant mortality. Social responses include increased aggression, infanticide and cannibalism, curtailed reproduction, abandonment of unweaned infants, and a rising incidence of unusual and unproductive sexual behavior, especially homosexuality and pedophilia.

Some researchers have suggested that this general adaptive syndrome may in fact play a far more important role than the availability of food in the regulation of mammal plagues, since a curious feature of most plague collapses in the wild is the continued availability of food and the scarcity of malnourished corpses.[14] The evolutionary value of such a mechanism would be considerable. If a plague species could cull its numbers before it totally exhausted its food resources, then it would avoid the risk of degrading its environment to the point that its extinction was guaranteed.

Uninhibited by most of the estrous or seasonal factors that regulate reproduction rates in other mammal species, *Homo sapiens* seemed, until recently, to exhibit no trace of this syndrome and thus appeared to be biologically defenseless against exponential population growth. Wherever overcrowding coincided with high mortality rates (due to disease and malnutrition), fertility tended to accelerate, as parents tried to ensure against possible future family losses. Consequently, because of cultural factors and a peculiar ability to plan ahead, humans have tended to reproduce fastest at precisely those times when the general adaptive syndrome might have been expected to come into play.

The Stress Connection

That appeared to be the case until the 1970s, when the East German scientist Gunter Dörner heard of the work of Selye and others on the physiological effects of social stress in laboratory rats and mice. Dörner had already spent much of his career investigating the role hormones such as estrogen and testosterone play in the development of the fetus, particularly in determining an individual's sexual orientation. He was especially intrigued by reports that female rats subjected to severe social stress during pregnancy tended to produce male offspring that were attracted only to other male rats. Since he already knew that high stress levels experienced during pregnancy resulted in lower levels of male hormones in the womb, he decided to survey that section of the German population that had been born during and just after World War II to see if he could detect statistical evidence of the link between stress and sexuality in society at large. From the eight hundred homosexuals questioned in the survey, Dörner was able to determine that the homosexual birthrate (i.e., the percentage of homosexuals in the overall population) had indeed fluctuated according to the levels of stress suffered by women during pregnancy. More homosexuals had been born during the years of greatest social stress—during the last months of the conflict and those immediately following the war—than at any other time. In fact, two-thirds of the homosexual men and their mothers reported prebirth levels of stress ranging from "moderate" to "extreme," and the stated causes included bombing, rape, and extreme anxiety. By contrast, only 10% of the heterosexuals in the control group reported unusual maternal stress during the prenatal period, and the mothers of these individuals reported only moderate prenatal stress.[15]

Such a rise in the homosexual birthrate would of course have had virtually no effect on the fertility potential of the German population as a whole, but that such a stress-related reproductive control mechanism existed in *Homo sapiens* at all was significant. The global fertility decline that has developed so unexpectedly during the past two decades clearly suggests that we may, after all, be subject to a multitude of subtle, large-scale fertility controls very like those that control rodent plagues. All the hormonal signatures are there, although it is impossible to discern at this early stage which forms of stress are primarily responsible for the decline.

Changes in lifestyle, self-administered recreational drugs and food additives, increased exposure to chemical pollutants that imitate or inhibit the body's sex hormones—any or all of these, plus others as yet undetected, might be the trigger. We will no doubt continue to equate our declining fertility with cultural success and attribute it to the empowerment of women, effective sex-education programs, and the general spread of literacy and better information. But if you switch off the cultural sound track and concentrate on the figures and the population graph, it all looks astonishingly like the end of a typical mammal plague—and the beginning of Selye's general adaptive syndrome.

The Enemy Within

Hormones move in mysterious ways their wonders to perform. As the body's chemical messengers they bring about adaptive physiological responses to environmental change. How elegantly appropriate it would be if, in chaotic and fractal fashion, they served precisely the same purpose on a global scale. Even manufactured substances that mimic hormones, a rogue element as far as we are concerned, may effect adaptive physiological changes. These substances, often called "endocrine disrupters," disturb normal sexual development in a wide variety of vertebrates, including humans, causing lower sperm counts, undeveloped or malformed genitalia, repeated failure of embryo implantation, and a rising incidence of ectopic pregnancies. They occur in a vast range of commercial products—pesticides, plastics, paints, inks, industrial detergents—and are released as breakdown products. The onslaught of such substances was first brought to public attention in 1962 by Rachel Carson in her momentous book *Silent Spring*. However, she considered only the toxicity of such chemicals and was unaware of their even more serious long-term hormonal consequences.[16]

Most of the known offenders imitate estrogen, and the impact is therefore most noticeable in males. Under the feminizing influence of these estrogen mimics, genes that are responsible for the production of testosterone in males often fail to switch on. Extensive surveys of male fertility carried out in several developed countries (notably Denmark, Scotland, and France) suggest that the average number of sperm in the ejaculate of a normal man appears to have fallen by up to 50% during the past fifty years. Although the method of analysis used in the surveys was widely criticized, similar sperm count reductions and increasing signs of sexual dysfunction have since been detected in several other vertebrates, especially rats, alligators, and birds, and these too seem to have been caused by exposure to artificially produced hormone mimics. The matter was finally put beyond reasonable doubt in January 1997 when a Finnish team compared testicular tissue (taken during postmortem examinations) from men who had died suddenly either in 1981 or 1991. In that ten-year period the numbers of men who had normal testicular tissue and healthy sperm production had fallen by more than 50%.[17] As increasing numbers of artificial endocrine disrupters become interwoven with whatever endogenous fertility inhibitors constitute the human version of the general adaptive syndrome, it seems likely that the current fertility decline will gradually accelerate.

Driven by such subtle and complex chemistry, our ultimate population decline is likely to remain both inexplicable and beyond remedy. We will see a gradual increase in the incidence of biological and social dysfunction, as well as rising levels of unproductive sexual behavior, such as homosexuality, lesbianism, and pedophilia, and an increasing heterosexual tendency to postpone or avoid parenthood. These natural fertility brakes are likely to become even more pervasive and pronounced as population pressure increases, social cohesion decays, and environmental degradation mounts.

Although the global fertility rate is now showing signs of significant decline, judging by the current rate of industrialization, our energy consumption will multiply by a factor of 2.5 by the year 2050.[18] The technology factor of the I = PAT equation (Impact = Population × Activity × Technology) thereby promises to compensate for most of the decline in population growth. In other words, as far as the biosphere is concerned, it is almost as though human fertility were not declining at all, suggesting that something other than simple math is manipulating the equation to achieve this convenient end. To lay the blame on human culture is circular reasoning and provides no answer, so we are left facing the intriguing alternative that our predicament betrays the presence of some form of evolutionary genetic management system, a mechanism that has automatically locked us into a cultural development that guarantees disaster; perhaps even a Gaian mechanism like the one proposed by Vernadsky, Lovelock, and Margulis.

Family Traits

The long reproductive window and high fertility rates that drove our population juggernaut to this evolutionary precipice seem both bizarre and unique when compared to the sexual norms of most other modern primates. However, one very persuasive piece of evidence suggests that our peculiar brand of sexuality may in fact reflect a time-honored family characteristic. The closest of all our animal relatives, the diminutive bonobo

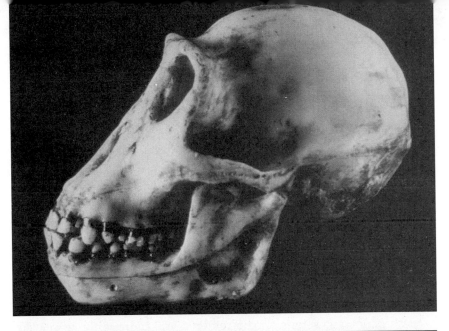

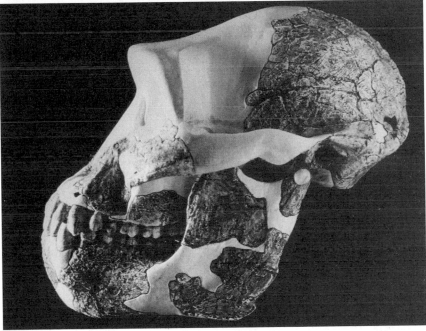

Bonobo skull (*Pan paniscus*), top, and "Ricky" (*Australopithecus afarensis*), bottom. The close relationship between humans and bonobos is clearly displayed in the profiles of these skulls. On the top is the skull of a modern bonobo; the reconstructed skull on the bottom is 3.2 million years old and belonged to an australopithecine male affectionately known as "Ricky." Australopithecines are believed to be directly ancestral to humans. (Bonobo skull and cast of "Ricky" courtesy of the Cleveland Museum of Natural History.)

(*Pan paniscus*), even to humans appears to be obsessively sexual. Sexual interaction of one kind or another appears to mediate almost every bonobo interaction from the cradle to the grave. It breaks the ice during introductions, serves as the appetizer to most meals, helps to pass the time between meals, and provides the currency that oils the wheels of most other social intercourse. It is used to defuse tension between individuals and between groups, and it is even employed to mollify and comfort fractious children. Bonobo society is consequently remarkably peaceful. All members know each other rather well, regularly making love, not war; disagreements are rare. Although it may sound as though bonobos are constantly "at it," they are not. Sexual encounters may occur with astonishing frequency, but for the most part they are not only casual and commercial but also remarkably brief. Even with full penetration a mating usually lasts only about thirteen seconds.[19]

Two other startling features serve to distinguish bonobos from their more robust and rambunctious cousin, the common chimp (*Pan troglodytes*). Bonobo society is female-centered and thoroughly egalitarian. And perhaps even more disconcerting to their human observers, not only do bonobos frequently indulge in face-to-face intercourse, but they also engage in other forms of sexual gratification—manual, oral, and genital—and in almost every possible partner combination available. In fact, sexual contact between parents and offspring ceases only when the young reach sexual maturity and they would run the risk of producing inbred offspring. Even the finding of food provides an excuse for sexual dalliance. Then, when food-sharing negotiations begin, sex seals the deal. Not only do males and females in the wild indulge themselves at the drop of a banana, but captive males often become sexually aroused merely by the sight of a fruit-bearing zookeeper.[20]

Female bonobos, like their human counterparts, have been structurally modified to make the most of these frequent and socially useful sexual exchanges. As in humans, the female's vulva and clitoris are much more frontally oriented than are those of the common chimp. She is almost continuously sexually attractive, even during lactation, and it is often the female that initiates sexual contact. If her partner happens to be male, then it is three-to-one she will present him with the missionary position.[21]

Their nonsexual behavior similarly confirms their close genetic link with humans. They stand upright more often than *P. troglodytes,* and they walk with a grace that sets them apart from their waddling cousins—a surprising feature considering that bonobos, unlike *P. troglodytes,* never left the shelter of the forest and therefore never had to walk very far. Since, in these and other features, bonobos seem even more similar to our bipedal australopithecine ancestors than they are to the common chimp, and since they lack some of the anatomical specialization of the common chimp, it seems likely that bonobos represent a genetic centerline from which humans and *P. troglodytes* diverged. According to de Waal, it is likely that the bonobo has undergone "less transformation than either humans or chimpanzees, and most closely resembles the common ancestor of all three modern species."[22]

The Territorial Imperative

A major difference between ourselves and bonobos is that they rarely eat meat, and then only when small mammals, such as rodents or baby monkeys, fall easily into their hands. With a distribution that has always centered on Africa's tropical rain forests, bonobos would have rarely run short of their preferred diet of fruits and would never have been forced to hunt for additional protein as were our ancestors. Consequently, bonobo territoriality, unlike ours, is low-key and causes little conflict. Serious aggression between neighboring groups has been witnessed in the field, but it is rare, and peaceful mingling is the rule, with grooming and sexual activity serving to break the ice between strangers of both genders. In fact, sex seems to be the glue that binds bonobos society together, defining social status and minimizing conflict, both within the tribal group and without. This suggests that if humans did evolve from a bonobo-like ancestor, our aggression and territoriality must have been subsequently developed by the harsh constraints of life on the drought-ridden plains that lay to the east of Africa's Great Rift Valley.

Bonobo distribution is still restricted to central Africa's humid forests, and the entire population now seems to be confined to a region south of the Zaire River where perhaps fewer than 10,000 survive—a surprisingly small number in view of their extraordinary promiscuity. But reproduction is a slow business for them. A female bonobo in the wild does not usually give birth to her first child until she is about thirteen or fourteen. She then nurses and carries this solitary offspring about with her for some four or five years, during which time she remains sexually unavailable. She is therefore able to conceive only once every four to six years, allowing her a maximum of some five or six progeny throughout her forty-year life.[23] The human female, in dramatic contrast, has a slightly longer reproductive life that allows her the time and biological opportunity to bear at least twenty children. Allied to our bonobo-like appetite for sex and our ability to feed and service large populations, this drastic increase in human fertility gives the human species real plague potential.

A plague animal is ultimately defined by the graph of its fluctuating population; if there is one thing we share with species like the house mouse and the plague rat, it is the shape of the spike that has recently appeared on our population graph. Such plague spikes display the feature Malthus noted in 1798—exponential growth, which occurs where the rate of increase is restricted only by the number of individuals, their net reproductive rate, and the existence of adequate resources. On a finite planet, having adequate resources is necessarily a temporary state of affairs. Consequently, the graph of any animal plague is characterized by a gently curving foot which rises ever more steeply to a sharply defined peak. On the far side of that peak, the reverse occurs. The line plunges precipitously before flattening out at the bottom as the population returns to a sustainable base figure. It is this collapse curve that should most concern us now, particularly in view of the recent decrease in human fertility.

Exponential Growth

The explosive nature of exponential growth was graphically illustrated during an international conference at Australia's Academy of Science in Canberra some years ago. In our intestines we harbor a very useful form of one of the world's best-known bacteria, *Escherichia coli*. In thirty minutes one invisibly small bacterium becomes two bacteria, and in the next thirty minutes, four bacteria. With the passage of an hour, although the bacterial mass has quadrupled, the threat is not yet visible to the naked eye. In two hours, with sixteen bacteria, the same would still apply. Yet, if the process had begun at midnight and remained unconstrained by environmental factors, by 10:30 P.M. on the third day the bacterial mass would be equal to half the weight of the planet. Just thirty minutes later, if their reproductive rates remained steady, those bacteria would, in one cataclysmic convulsion, reproduce the other half of the planetary mass.[24] By the end of that day, our original organism would have produced a biomass equal to four earth-sized planets.

It would, of course, take us humans a little longer to match such a feat, but that is precisely the kind of statistical curve we have been riding for at least two centuries. About 6 billion of us now live on the planet, and although the global birthrate has eased significantly during the past decade, roughly 80 million additional individuals are born every year.[25]

Our population curve has followed the typical plague profile exactly. Provided that some unforeseen factor does not interfere with our fertility rate in the next thirty to forty years, the population should peak somewhere between 7 and 7.5 billion and then begin the second phase of its plague cycle, the collapse. And if the human plague really is as normal as it looks, then the collapse curve should mirror the population growth curve. This means that the bulk of the collapse will take not much more than one hundred years, and by the year 2150 the biosphere should be safely back to its preplague population of *Homo sapiens*—somewhere between 0.5 and 1 billion. (That is, of course, if we are not altogether extinct.) The typical processes of evolution will then resume and, as in Lovelock's Daisyworld, the biosphere will begin restocking its gene pool, restoring biological diversity, and rebalancing its geochemical and biochemical systems.

Apocalypse Past

Although human culture seems to have integrated quite seamlessly with natural factors to produce a perfectly normal plague graph up to this point, nevertheless we cannot be certain that our pattern of collapse will conform as precisely. It might be argued that certain forms of stress seem to bring out the best in us, and judging by the fossil record, environmental stress may have been a key element in our original evolutionary success. However, the environmental stresses that threaten to engulf us in the future are significantly different from those on which we sharpened our wits 2 million years ago. And by the middle decades of the twenty-first century they will constitute only half of our problems: famine, disease, and social disintegration will provide the other half. The evidence that these three factors have been our nemeses many times in the past lies scattered around the world in the ruins of cultures that were once rich and vibrant, including some of the world's oldest civilizations: Sumer, Egypt, Mycenae, Petra, and the Harappan culture of India's Indus valley, as well as the early American civilizations of the Mayas, the Anasazi, and the Aztecs.[26]

The sequence of decline is now well documented. Prolonged cultural success and population growth depends on progressive deforestation and agricultural expansion. The decreasing vegetation cover alters the microclimate, producing unreliable rainfall and increased erosion by wind and water. Good soils disappear, harvests decline, and famines cause growing social unrest and an ever increasing reliance on religion and public ritual as the society searches for solace and salvation. Consequently, the lavish temples, tombs, and monuments that usually accompany this late stage of cultural development are not symptoms of prosperity and cultural climax, as was once thought, but rather are the harbingers of doom.

Easter Island

Precisely this grim progression of events seems to have occurred very recently on Easter Island in the eastern Pacific Ocean. The crew of a Dutch vessel who found their way to this lonely volcanic island in 1722 were confronted by row after row of gigantic stone statues, some of them up to 10 meters (33 feet) tall. Yet almost no other trace was left of the civilization that quarried, carved, and erected these massive monuments. When Captain James Cook visited the island in 1774, he noted that there were fresh wounds on some of the islanders; the expedition's artist, William Hodges, included skeletal remains in one of his paintings. Conflict was clearly part of daily life for the stranded islanders. By this time, the island itself was virtually treeless and occupied by about 2,000 malnourished inhabitants, many of whom lived furtive lives in caves or crude boulder shelters; they scratched a meager living from the island's coastal fringe. Without wood, they could not build proper houses, nor could they build boats to use either for fishing or escape from bloodthirsty neighbors.[27]

The history of the island's people and their stone statues—between 800 and 1,000 of them—remained a mystery for more than three centuries, until careful reexamination of the few remaining cultural artifacts and detailed analysis of the island's well-preserved pollen record gradually revealed the grim record of cultural boom and bust. It appears that Polynesian explorers discovered and settled the island somewhere between 1,400 and 1,600 years ago, giving birth to one of the most vibrant, literate, and artistic cultures in the entire Pacific region.[28] But that cultural success was their undoing. The population grew steadily until about 900 years ago and then went into exponential overdrive, reaching a peak of about 10,000 people in the late seventeenth century. Having undermined its own environmental foundations, this potently mystical culture then proceeded to tear itself apart, all within the space of about one hundred years. Their downfall appears to have been their own vigor, ingenuity, and spirituality, allied to shortsighted agricultural practices. Once the last of the island's lush palm forests had been felled and all the arable land had been turned over to farming to feed the burgeoning population, they could no longer build the canoes that previously enabled them to fish and would have allowed some of them to escape from their island prison. Meanwhile Polynesian rats, introduced by their ancestors, ate the palm seeds that might have allowed reforestation. Their fate was now sealed.

With the island's tree cover permanently removed, rainfall declined, erosion increased, harvests shrank, and famine befell the unwitting inhabitants. With desperation feeding their mysticism, the islanders seem to have neglected the farming and beach fishing that might have fed them and turned with renewed vigor and dedication to carving and erecting their statues. The most massive of these still lies unfinished where it was quarried. It is 20 meters (65 feet) long, weighs about 270 metric tons, and appears to have been too heavy to move, even for these experienced stonemasons. Dividing finally into two armed camps, the islanders began to fight continually with one another. Culled by starvation, infighting, and the depredations of visiting slave traders, they almost achieved their own extinction. By 1877 only 111 remained.[29]

In retrospect, the fate of the Easter Islanders seems to have been decided from the moment they first set foot ashore, for there seems to have been a grim inevitability to the tragedy that unfolded there. With no room to expand their food resources and no avenue of escape, their tiny paradise soon became a prison, and on a smaller scale they faced precisely the scenario that now confronts modern humanity. Lest this example be shrugged off as a solitary aberration, let us examine one other case.

The Maltese Malaise

New archaeological excavations on the Island of Malta in the Mediterranean suggest that a disturbingly similar progression of events occurred there more than 4,000 years ago. Four factors were common to both cultures, and it appears to have been these that locked them into a similar trajectory from prosperity and abundance to disaster. Those factors were their island's initial fertility and isolation, their vigor and ingenuity, and ultimately their obsessive mysticism.[30]

The Maltese Islands today are rocky and barren with little vegetation and less water, yet the geological evidence suggests that 6,000 to 7,000 years ago, when the islands were first colonized, the landscape was quite different. In fact it was abundant timber, good water, and fertile soils that launched Malta's first settlers on the path to cultural success, and for a thousand years the population grew and prospered. But by 4,000 years ago the islands' natural assets were gone and its people were doomed. Once they had removed the trees in order to farm the land, rainfall declined, erosion robbed them of soil, and abundance shriveled to subsistence.

What finally became of the people is not known. As at Easter Island, all that is left of them is the massive evidence of their obsessive mysticism: an astonishing array of temple ruins clustered into about twenty groups, each of which includes at least two or three massive stone structures, and a series of extensive underground burial chambers. According to a team of British and Maltese archaeologists who recently reexcavated the site, radiocarbon dating has unequivocally assigned all of the temples to the latter stages of the culture—after the best of the soil had blown away and hardship had set in.[31]

Between about 6,000 and 5,500 years ago, the caves and underground tombs, like the grave artifacts they contained, had been relatively simple and undecorated, paralleling those of a contemporary culture in Sicily.

Significantly perhaps, the abundance of carved figurines depicting obese or pregnant females suggests that the society was preoccupied with fertility very early in its growth. About five hundred years later, however, the culture underwent a dramatic change, and a cult of the dead appears to have taken over. Old burial sites, originally dedicated to small, simple family funerals, appear to have been commandeered by a powerful priesthood who enlarged and redesigned them to accommodate elaborate death ceremonies. At some sites the main chamber was furnished with a semicircle of massive stone slabs arranged about a large stone bowl, with rows of burial compartments carved into the rock walls behind them. One of the largest of these burial sites housed the remains of between 6,000 and 7,000 individuals, while another incorporated a huge burial pit that served either as a mass grave or a bone repository.

Artifacts from this period consistently depict obesity in humans and animals and also include numerous phallic symbols, either carved in stone or bone, or modeled in clay, pointing to an obsession with the living world and its successful propagation. According to the archaeological team, "vast amounts of human time and energy were invested in temple building, artistic endeavours and ritual feasts." Meanwhile, "the people seem to have expended relatively little effort on the building of villages or domestic structures, on terracing or on developing better farming techniques. The obsession with the cults of the temples seems to have been complete."[32] In fact, there are so many major temples scattered throughout this relatively small island group that they indicate a fierce religious rivalry between their builders.

Finally, the consistency of the skull architecture displayed by human remains at all the sites shows that there was little or no change in the genetic makeup of the people throughout their occupancy. In other words, this community remained relatively isolated throughout its existence and appears to have died merely from overpopulation and a surfeit of its own cultural success. The main reason the same kind of total collapse did not befall all early civilizations seems to be that they were not all physically isolated and people could usually move elsewhere when living conditions worsened. If Malta and Easter Island do indeed represent our earthbound civilization on a smaller scale, then the prognosis is not good.

These examples of environmentally triggered cultural collapse have been revealed only recently as new techniques of physical and statistical analysis have enabled archeologists to reassess in molecular detail the slender fossil clues that linger on in the ruins of ancient civilizations. Circumstantial though such evidence may be, it is sufficiently characteristic to alter the way we view culture itself. No longer can we pretend that culture is simply a priceless accumulation of history, wisdom, and technology that passes from generation to generation. Culture also contains a heady brew of mystical delusions that seems specifically designed to prevent us from gaining a stranglehold on evolution. Where a powerful culture threatens to overwhelm its natural resources, the magical potion of mysticism comes to the biosphere's rescue, intoxicating and immobilizing the offenders so that the environment can step in and deliver the coup de grâce. In this sense, then, it seems that our culture bears with it the antidote to evolution's most dangerous innovation—human intelligence.

Chapter 7

The Terminators

Who is the slayer?
Who is the victim?
—Sophocles

Gastric-Brooding Frogs

High in the mountain rain forests of eastern Queensland, there once lived two unique species of frog. Although the record of their existence is unequivocal, their brief, spectacular appearance in the scientific literature, their bizarre life cycle, and their mysterious disappearance lend their story a fairy-tale quality. They alone had managed to leap the taxonomic boundary between amphibian and mammal during reproduction by converting their standard vertebrate stomach into a cross between a mammalian uterus and a marsupial brood pouch. This unique strategy gave their offspring unparalleled security during hatching and brooding and afforded the young frogs a trouble-free birth through the mother's mouth.

The reproductive strategies of frogs are among the most diverse in the animal kingdom. The males of one species of Australian frog carry their young marsupial-style, in well-tailored "hip pockets," while the young of a South American species take refuge in their fathers' vocal sacs until they are old enough to survive on their own. It is as though evolution tried every trick in the book to match the highly successful mammalian combination of intromission and uterine development. But despite all that amphibian diversity, gastric brooding stands out as the most innovative and ambitious experiment of all.

Sadly, however, both varieties of gastric-brooding frog now appear to be extinct. Discovered only by chance in 1972, the first of these fairy-tale animals, *Rheobatrachus silus*, had vanished within eight years. In fact, the last sighting in the wild was in 1980, although the last captive individual did not die until 1983. Barely two months later a second species, *Rheobatrachus vitellinus*, was discovered in a similar mountain habitat some 800 kilometers (500 miles) farther north.

Although closely related and outwardly similar, each species achieved its reproductive transformation from amphibian to simulated mammal by slightly different means. Once inside their mother's stomach, the eggs of *R. silus* avoided being digested by shutting down the particular cells that produced their mother's primary digestive agent, hydrochloric acid. These acid-producing cells then remained inactive throughout the tadpole stage of development until the fully formed young frogs were ready to emerge into the world.[1] Clearly, the ability of the *R. silus* embryos to halting the production of their mother's digestive juices held great promise for millions of humans suffering from hyperacidity and gastric ulcers.

The offspring of *R. vitellinus* achieved their immunity from digestion by covering themselves with a mucous coating that was impervious to hydrochloric acid. In neither species could the mother feed during her "pregnancy," a period that could last up to eight weeks. In fact, the stomach became so distended in *R. silus* females that it partially detached from the surrounding tissue. It also compressed the lungs to the point that the mother could no longer breathe and had to rely on respiration through her skin.[2]

Birth by Mouth

During the short time the frogs were available for study, several "births" were observed. The first opportunity for such observation came as a complete surprise. An *R. silus* female was being moved from one tank to another and became highly stressed. She then rose to the surface, opened her mouth, and ejected a clutch of six young frogs by propulsive vomiting. Since almost nothing was known about the frog, researchers delightedly concluded that they had discovered an Australian version of a South American species in which the male carries the young in its vocal sac. The true magnitude of the find was not revealed until frog expert Mike Tyler

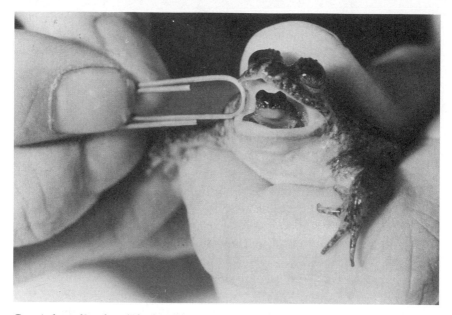

Gastric-brooding frog (*Rheobatrachus silus*). Pictured at the moment of its unique birth, a young *Rheobatrachus silus* frog pauses inside its mother's mouth before launching itself into the world—in this case, into oblivion. Neither species of gastric-brooding frog has been seen since 1985; they are presumed extinct. (Photo courtesy of Mike Tyler.)

of Adelaide University dissected a second postnatal parent and discovered not a male with an enlarged vocal sac but a female with an inactive but hugely distended stomach.

Most birthing events were much more leisurely affairs than occurred in the first two cases. In one instance a female "gave birth" twenty-six times over a period of seven days. Before each birth the mother rose to the surface of the water and opened her mouth very wide; one or two young frogs then appeared on her tongue, paused for a moment, and casually hopped out.[3] Similar behavior was later seen in *R. vitellinus*. By 1985, however, *R. vitellinus* had gone the way of *R. silus*, taking most of its evolutionary secrets with it. While both had been seen giving birth, neither had been seen laying eggs or ingesting either eggs or young, and almost nothing was known of their behavior in the wild.

Although extensive logging and some gold panning may have contributed to the demise of *R. silus*, Mike Tyler believes that the loss of both species is simply part of the wave of frog extinctions that is currently sweeping the world. One plausible theory suggests that since frogs, being amphibians, are able to respire through their skin, most tend to be more sensitive to air and water pollutants than are reptiles and mammals. The theory goes on to suggest that although the pollutants may not be fatal in themselves, they lower the frogs' immunity to a wide variety of diseases that would not otherwise kill them. Given their singular vulnerability to substances that ultimately could threaten the biosphere, frogs may serve as the miners' canary in that their demise provides an early warning of dangerous environmental degradation. Most of the species on the casualty list have never even been seen, however, let alone investigated. They are simply part of the inevitable collateral damage humans inflict on the biosphere in their pursuit of wealth and personal satisfaction—or mere subsistence in many cases.

Nearly half of humanity's medicines are drawn from, or based on, natural ingredients extracted from the very few species with which we are passably acquainted. Of the world's higher plants, for example, scientists have screened only 0.5%, and these now provide the bases for forty-seven of the world's major pharmaceutical drugs. Yet according to a recent survey by an American research team, tropical forests contain about half the world's 125,000 species of flowering plants, and each plant will yield an average of six compounds that have medicinal potential. It represents undiscovered drugs worth at least \$147 billion, calculated at current value-to-society rates.[4] Nevertheless, the world's tropical forest, already reduced to half its preindustrial size, is disappearing faster than ever. There is even good reason to suspect that the global extinction rate among all species may be directly linked to the growth rate of humanity's environmental impact via the I = PAT equation (Impact = Population × Activity × Technology) outlined in chapter 2, and that the rate of species loss is therefore growing exponentially. What can be said with greater certainty, though, is that species are currently disappearing more than 1,000 times faster than they can be replaced by genetic evolution. To put it bluntly, earth's biota is presently undergoing a mass extinction of awesome proportions, and we, as the primary cause, cannot hope to escape the consequences of this massive genetic hemorrhage.

Maori Bird Massacre

Although the anthropogenic extinction of any species, plant or animal, is clearly to be mourned for its own sake, in the sense that the extinction of potentially supportive species tends to loosen our grip on the planet as a whole, they also represent a tiny but vital cog in the Gaian machinery of planetary management.

The link between human activity and species loss is not just a twentieth-century phenomenon, of course. The fossil record shows that the arrival of human beings in an area has always coincided with a wave of extinctions, invariably among the larger, "prime target" species, the ice-age megafauna. In New Zealand, for example, there is now little doubt that the Maoris, in something like five hundred years, totally exterminated the giant flightless moa, as well as many other large flightless birds, and one very aggressive species of gigantic eagle. The eagle used to prey successfully on the largest moas—and, no doubt, on Maoris when the opportunity arose—but it never approached the killing efficiency of human predators. At one campsite on the South Island, researchers uncovered the bones of about 9,000 moas and the remains of 2,400 moa eggs, while at another South Island site the mass of bone fragments suggests that somewhere between 30,000 and 90,000 moas had been slaughtered and eaten there. These huge bony rubbish dumps are all that remains of the most astonishing bird fauna of the modern world, the remnants of a unique evolutionary development that took place because the birds had no terrestrial predators. The moas, for example, ranged from 3-foot-high miniatures to 500-pound monsters that stood 10 feet tall. The first Maoris arrived in New Zealand somewhere between 1,000 and 800 years ago, and by the time the first Europeans landed there in the seventeenth century, the Maoris had killed between 100,000 and 500,000 moas, driving all twelve species to extinction, along with the giant swans, coots, ducks, and eagle.[5]

Australian Extinctions

The same story has been repeated countless times around the world, most notably on islands such as Madagascar (elephant birds, giant lemurs, giant tortoises, and giant "pygmy hippos"), Mauritius, Reunion, and Rodrigues (the dodo). In Australia, where once there had been a wide variety of giant marsupial herbivores and predators, only the medium and small species of herbivores remain, and most of the losses are attributable, indirectly at least, to the impact of expert human hunters. These hunters' frequent use of fire to flush out game and promote new growth eventually killed off most of the fire-sensitive vegetation throughout the continent. The result was not only a dramatic reduction in the total volume of vegetation and a lethal loss of habitat for many species, but in all probability it also contributed significantly to the regional climate change that was already in progress. It is now widely believed that this initial environmental degradation, compounded by the efficiency of the new hunters, contributed to the extermination of most of Australia's larger marsupials, animals like the hippo-sized *Diprotodon optatum*, several giant browsing kangaroos, flightless birds almost as big as the moa, and a leopard-sized arboreal carnivore, *Thylacoleo carnifex*. There is no evidence of traditional butchering sites like those in New Zealand, but the fire-assisted environ-

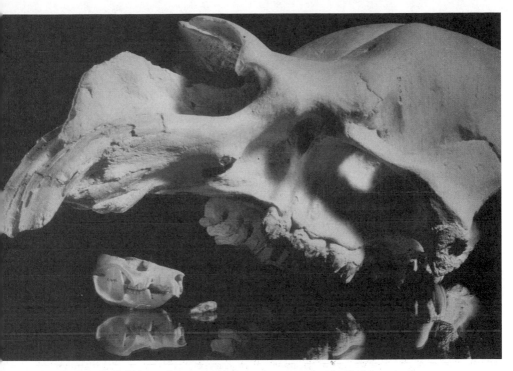

Diprotodon skull (*Diprotodon optatum*). Looming over the skull of two modern marsupials, a possum (left) and a dunnart (right), is the skull of a rhino-sized marsupial known as a diprotodon. The adults were the largest marsupials that ever lived. The skull in the photograph belonged to a juvenile. (Skulls courtesy of the National Museum of Victoria.)

mental changes would have guaranteed the extinction of large herbivores. There is also reason to believe that the extinction of this megafauna itself contributed to further climate change and desertification by the loss of the major recycling mechanism they provided in their droppings. According to Tim Flannery of the Australian Museum, Sydney, a global comparison of genera extinctions shows that Africa (south of the Sahara) lost only 5%, Europe lost 29%, and North America lost about 73%. Australia, by contrast, tops the extinction list with a loss of 94% of its megafaunal genera.[6]

It is the lack of synchronicity between these extinctions around the world that best betrays the cause. Whereas Australia's megafaunal extinctions began more than 40,000 years ago, New Zealand's seem to have been crammed into the past 800 years. The American losses are more equivocal, however, coinciding not only with the arrival of humans but also with the dramatic climate change that marked the end of the last glacial era 12,000 years ago. Although few of the extinctions may have been primarily caused by humans, human hunting would have contributed in many cases.

The Price of Progress

Those early waves of extinctions were only the beginning, however, especially in Australia. In fact, the general process of environmental degradation was probably begun by Australia's archaic hunter-gatherers as much as 120,000 years ago and was then continued by the Aborigines when they arrived some 60,000 years later. It had largely come to a halt by the time the first European settlers arrived in 1788, but the subsequent introduction by Europeans of hard-hoofed grazing animals into an environment already compromised by Aboriginal occupation ensured that large areas of Australia soon began to slide into a whole new cycle of degradation. The environmental onslaught was compounded by the gradual removal of some 70% of the native vegetation, the introduction of at least eighteen exotic mammals into the wild, and the growth of huge feral populations of five of the most damaging species—cats, foxes, rabbits, pigs, and goats.[7] These factors combined to produce the extinction of at least another twenty mammal species, or one-third of the world's total mammal losses for the past five hundred years.

Continued loss of vegetation and topsoil eventually renders any region uninhabitable to most species. About 55% of Australia's agricultural land has now been degraded, in some cases beyond redemption, by the desertification process.[8] This dismal record of environmental degradation represents an astonishing impact by a tiny population on a landmass that is almost the size of the United States and graphically illustrates that for an animal of our size we tread with massive ecological weight on the vast multitude of organisms that share this planet with us.

Environmental Instability

A number of our catchphrases indicate the nature of one of our more serious mental handicaps: "The more things change the more they stay the same"; "History repeats itself"; "It will all be the same a hundred years from now." How could we have been so blind, even when professional mystics like the Reverends Malthus and Lütken predicted our problems with such clarity two hundred years ago? Certainly the natural world does not concur with our clichés, and never did, as the fossil record clearly shows. Global temperatures and climates have fluctuated wildly throughout geological time. As recently as 18,000 years ago the mean global temperature was about 5°C cooler than it is now, and the seas were at least 100 meters below their present level. Unfortunately, all those dramatic changes occurred before the birth of civilization, and relative climatic sta-

Queensland lungfish (*Neoceratodus forsteri*). An extraordinary exception to the general rule of extinction, the Queensland lungfish has altered little in 370 million years. The Australian species is the most primitive of the three extant groups and represents the oldest surviving genus—and species—of the animal world.

bility is all we have known in a historical sense. It might justify our delusion, but that delusion has left us spectacularly vulnerable.

Careful analysis of the fossil record shows, in fact, that major environmental traumas have often beset the planet, slashing its biota to ribbons on several occasions. Far removed and irrelevant though these events might seem, such massive evolutionary blows are a key ingredient in the recipe for a biologically healthy planet. Evolution as we know it could not have occurred without them. Ever since the very first organic molecules appeared, the incessant conflict between the relatively stable chemistry of life and an inherently unstable environment has produced an astonishing spectrum of genetic variation. This is the raw material that feeds the machinery of Darwinian selection with sufficient candidates to ensure that at least a few survive each time the environmental rules change.

On the other hand, if one or two species become so successful that they blitz the opposition, leaving fewer candidates for the selection process, then the viability of all life is undermined and the odds in favor of total extinction during the next environmental crisis increase significantly. This was the scenario when oxygen-producing bacteria exploded in the world's oceans more than 2.5 billion years ago, threatening the survival of all life and setting evolution on a new course. For evolution to proceed, therefore, not only must every genetic variant contain an evolutionary "escape clause" in the form of an adaptive asset that locks it to a particular environment, but the more successful and fertile the species, the more vital is this codified recipe for eventual self-destruction.

Such built-in obsolescence is normally provided by the specific nature of the organism. Custom-designed features such as heavy shells, short legs, or bright coloring might be decisive survival factors in one environment, but they also lock their owners to a narrow range of habitats and behaviors. When environmental change alters the rules for productive existence, such assets usually turn into handicaps, nudging the species into the fast lane to extinction. Most species are lucky if they last much more than a million years.

Our Stable Genes

Where human behavior is concerned, those old stability clichés are very well founded indeed. It is an entirely safe bet that the underlying thought patterns of the modern city dweller are virtually identical to those of our cave-dwelling ancestors more than 150,000 years ago when the modern version of our species first appeared. The reason is simple enough. Evolution is not, as is commonly perceived, primarily concerned with producing change, but rather with maintaining relative stability in unstable surroundings. Genes are the biological rocks in the stormy environmental sea. During the process of replication, DNA routinely achieves a level of accuracy undreamed of by human beings. Moreover, the process of replication is so complex and demanding that those errors that do manage to elude correction and become expressed in any meaningful way are almost invariably lethal, meaning that whenever a regional environment remains relatively stable, so do the genes that live there.[9]

Our global environment has not changed much during the past 10,000 years, and consequently, neither have we. However, one very important qualifier needs to be added here. It was between 12,000 and 10,000 years ago that human culture first divided into its two main streams: semi-nomadic hunter-gathers and those who chose to establish permanent, agriculture-based settlements. From that moment onward, slightly different cultural pressures began to operate selectively on the two genetic streams, and these pressures ultimately produced slight differences in the behavioral bias of each. In other words, the different behavior patterns that separate the modern hunter-gatherer from the rest of human society have absolutely nothing to do with so-called evolutionary progress. Both are equally advanced in evolutionary terms, but like identical tennis balls hit with an opposite spin, these two societies merely behave a little differently in flight. Left to themselves, each works well: transpose individuals from one culture to the other, and both struggle to survive.

It was the microevolutionary process of behavior selection that put the Aborigines so far ahead of the Europeans when they first arrived to settle in Australia two hundred years ago. An Aborigine could live very comfortably in regions where the survival of the average European colonist was—and still is—measured in days. The Aboriginal diet was far more diverse and nutritious, their society more structured and orderly, and by and large their lives were mystically rich, often leisurely, and generally rewarding. Levels of personal and intertribal aggression seem to have been much the same as those in European society, but their rigid systems of family and tribal law ensured that most disagreements were settled with minimal loss of life. What they did not have, however, was 10,000 years of acquisitive existence and technological development behind them—neither did they have firearms. When at last they were confronted by well-armed invaders from England, the outcome was never in doubt. The colonists looked on the Aborigines as less than human and consequently treated them as vermin. They were shot for sport, poisoned with gifts of arsenic-laced flour, and occasionally rounded up and slaughtered in military-style operations. Missionaries and educators tried to solve "the Aboriginal problem" with less violent methods, but all were ineffective.

Having failed to achieve a final solution with guns, arsenic, religion, or education, the colonial government then tried separating Aboriginal children from their "heathen" parents and "primitive culture" in the hope of assimilating them into "decent" European society.[10] Predictably, that too failed, leaving deep scars on the survivors. Inevitably, the Aboriginal population remains a disinherited fringe society, living for the most part in squalor and aimless discontent, plagued by unemployment, crime, poverty, disease, alcoholism, and domestic violence. Their plight highlights the behavioral chasm that inevitably exists between the descendants of all hunter-gatherers and the commerce-driven societies that have overwhelmed them. The most that can be said is that they survived. Most did not.

The Case against Us

Let us for a moment indulge our anthropocentric imagination and visualize ourselves facing a jury of cosmic peers. The prosecution is having a field day. "Here was a planet of moderate mass and gravity, perfectly placed in relation to the sun; a blue, rain-washed sphere, superbly managed by the diverse organic products of four billion years of painstaking biological evolution. Here, truly, was the Camelot of the cosmos. And look at it now! Its air and water polluted, half of its forests felled and burned, its wetlands drained, much of its topsoil blown into the sea, and the rich fabric of life that once adorned the biosphere torn to shreds. How could this galactic pearl have fallen so easily into such crude, unsuited hands? And what of the accused, these hairless, underdeveloped, quarrelsome apes, equipped with little more than glib tongues, sharp wits, and nimble fingers; how could they have wrought such havoc so swiftly? And on what authority?"

It sounds like a good prima facie case, and our claim that we are uniquely sentient and intelligent only incriminates us further. As for our defense, we seem unable to manage anything better than the old cultural excuse: "victims of a bad upbringing, Your Honor." Admittedly, traditional wisdom has it that humans are ultimately shaped by their culture, and all human cultures have been founded on the assumption that the earth's bounty is not only inexhaustible but divinely bequeathed. In fact, failure to believe in the principles of Growth and Progress represented not just economic heresy but a dereliction of moral duty. Yet if these excuses were to constitute our primary defense, surely we would be better advised to tell the truth and plead insanity.

Alternatively, we might run a Nuremberg defense on the basis—that as spiritual beings and good servants of our gods, we were "just obeying orders." But lacking proof of the hand that signed those orders, we would surely face a very rough cross examination. There is, however, an alternative version that might indeed hold up: "We are merely servants of our genes."

Built-in Flaws

Consider the genus *Homo* as it was some 2 million years ago. Here was an animal that had strayed a little too far from its ecological niche, and judging by its lack of adaptive structures, appeared to be earmarked for extinction. Yet, given its peculiar abilities to reason, communicate, and culturally modify its behavior, here also was a spectacularly versatile and resourceful survivor. But there was hidden danger, too, for here was a creature so cunning that it would eventually insulate itself from most of the natural shocks that other flesh was heir to: a creature able to appoint itself as executor of an evolution of its own devising, a culture-driven evolution that could both outrun and outgun biological systems in the day-to-day business of creating, modifying, and discarding other species. In short, here surely, was the most dangerous animal ever to walk the earth.

With the introduction of this wild card into the biospheric equation, evolution might have trumped its old adversary, the environment, in the short term, but unless some fatal flaw, some kind of concealed time bomb, could be installed in *Homo sapiens* to cap its explosive potential, there was the distinct possibility that 4 billion years of hard-won biodiversity might one day be sent down the cosmic drain with a mere flick of the sapient wrist. Since there is very real evolutionary value in a genetic mechanism that dictates self-destruction under such dire circumstances, there is good reason—and some evidence—to suggest that it does in fact exist. And since other plague-prone mammals seem to posses it, we should not presume we are exempt.

Finally, assuming a lethal device of this kind has been installed in *Homo sapiens,* where could it be concealed within such cunning creatures and still remain beyond the reach of their prodigious technology should they finally detect its presence? These are the questions to which we now turn.

PART III

Solutions

Chapter 8

The Spirit in the Gene

Chaos of thought and passion,
all confused
—Alexander Pope

The First Toolmaker

A traditional way to define humanity is to describe our species as the only toolmaking animal in the world. Other animals may use tools, we say, but none manufacture them—except in the most marginal sense. Our clever evolutionary siblings the chimps, for example, are known to crack nuts by crushing them with a rock and to strip leaves from saplings so that they can fish termites from their nests or honey from beehives. But neither selecting the rock nor obtaining a twig and stripping its leaves constitutes manufacture as we commonly understand it. Similarly, two great apes, an orangutan and the very sharp-witted bonobo named Kanzi, learned from their human handlers how to fracture rocks and then use the resulting sharp-edged stone chips to cut string that tied down the lids of boxes containing food. But both animals made their cutting tools by holding one rock in their hands and throwing or striking it against the other on the ground. Neither was able to progress to the human method of holding both stones and striking one against the other. So the quality of their stone-chip tools remained the product of chance, and our honor as rational beings remained intact.[1]

But one little-known toolmaker of extraordinary talent is never mentioned in this context. As night falls in my leafy Australian garden, a remarkable hunter stirs from her daytime hiding place, selects an ambush site, and begins to manufacture (in the truest sense of the word) a weapon that is elegant in design, precise in its craftsmanship, and lethally efficient in operation. This toolmaker is, of course, a spider. This particular variety of spider casts a net over its prey rather than weaving a stationary web. The weapon it constructs is so sophisticated that it took our ancestors more than 2 million years to match it. Its centerpiece, a small rectangular net fashioned from a knitted ribbon of specialized silk, is designed to be thrown over the spider's prey to immobilize it for the kill. The strategy is reminiscent of one that used to be employed by a class of Roman gladiator known as a *retiarius*, hence the generically common name, retiarius spider. Australia's retiarius spiders belong to the family Dinopidae, and their closest relatives live in Central and South America.

Retiarius Spiders

A Roman *retiarius*, wearing only a tunic and wielding a large rope net, was customarily matched against a fully armed swordsman. My small garden gladiators regularly attack almost anything that moves within their reach, a habit that occasionally pits them against such dangerous opponents as centipedes, scorpions, and other spiders. To capture such prey with least risk to themselves, they construct an aerial blind and make their net-casting lunge from there, as the victim passes through a carefully chosen target zone beneath them.

All retiarius spiders hunt in this fashion while they are juveniles, but as soon as they complete their final molt the males abandon their knitting and forgo all food in order to concentrate on sex. So by Christmas each year the adult females are the sole keepers of retarius technology. The smaller of the two genera known in Australia, *Menneus*, tends to hunt among broad-leafed foliage and mainly catches smaller ants foraging on the leaves. The larger Australian net-caster, *Dinopis*, needs larger prey and must stake out her hunting patch on the ground. And here the story gets very interesting indeed.

First she must select an appropriate ambush site. In the case of the two species of *Dinopis* that hunt in my garden, this means seeking a small patch of open flat ground fringed with tall grass or shrubs. Emerging from her daytime hiding place at twilight, she walks slowly about the area, tapping the ground with her front legs, apparently to detect twigs and other debris that might snag her net and spoil the attack. Having selected the

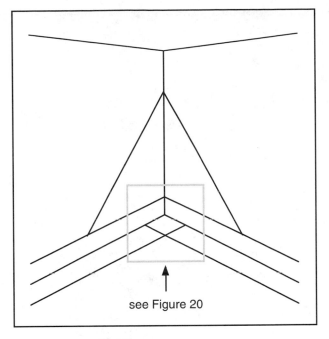

see Figure 20

Figure 19. Net-caster suspension lines, stage 1.

flattest, least-cluttered killing field, she then loops her primary silk threads (the first of three kinds in her arsenal) around two widely spaced and carefully chosen bollards on the ground and erects a complex triangular scaffold, attaching its apex to the foliage at the edge of the clearing (see figure 19). The trick is to ensure that this silken framework leans across the chosen target zone at an angle that allows her to position her aerial blind precisely over the target area and at the optimum height for a successful attack. Too low and the net may touch the ground and become snagged; too high and she may not be able to reach smaller prey at all. After adding some minor reinforcement to the structure, she then hangs head-up at the apex of the triangle and begins to knit her hunting net.

Weapons of Silk

Unlike the main scaffolding, the hunting net consists of all three kinds of silk, and she knits them together with her hind feet to form a silk ribbon of enormous strength and elasticity. The first two threads, providing strength, are like fine nylon fishing line. The next two threads are multifibered, like strands of wool, and the third kind of silk, extruded by a battery of glands known as a cribellum, looks just like a ribbon of shrink-wrap plastic. It con-

Net-caster knitting. Working the silk with hind feet linked together for added precision and support, a female *Dinopis subrufa* carefully manufactures the material for her hunting net by crimping the fine-fiber silk into minute loops and laying it on the main support lines. The process is so tiring that she alternates the crimping foot several times during the production of a single net.

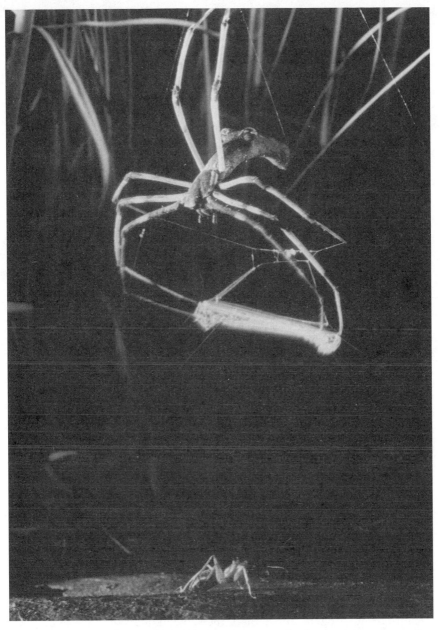

Net-caster attack. A *Dinopis bicornis* begins its net-casting lunge at an unsuspecting meat ant (*Camponotus* sp.).

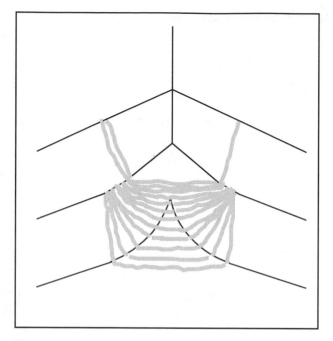

Figure 20. Net-caster hunting net, stage 2. (Densey Clyne, *The Australian Zoologist* 14:2 [1967].)

sists in fact of a parallel array of fibers so fine they can be distinguished only under a microscope. Although all of the silk emerges as a liquid compound of amino acids, it hardens on contact with the air, and none of the silk is equipped with the sticky coating that most garden spiders produce. The silken net of the retiarius spider snares its victims simply because its fibers are so fine that they catch on irregularities that are invisible to the human eye.

As the three kinds of silk extrude from the spider's silk glands, she lays one of her rear feet across them and with a tiny pushing motion crimps both of the "wool" strands and the ribbon of "shrink-wrap" into a continuous series of neat folds, at the same time laying these folds onto the heavy-duty threads that emerge much more slowly from her primary spinnerets. The three-layered ribbon of silk produced by this process is the key to the net's effectiveness. The two main threads supply a tensile strength comparable to steel wire of the same gauge, while the overlay of crimped fibers gives the net enormous elasticity and the ability to snag on the tiniest protuberance in any surface that it touches.

As soon as the small rectangular casting net is complete, the spider switches off all but her main spinnerets and performs the final and most astonishing step of the whole manufacturing process. She climbs to the apex of the scaffolding immediately above the hunting net and bites through the crucial suspension point that has until now kept the net under tension in the vertical position. This allows the net to fall into loose horizontal folds, suspended only by the side stays (see figure 20). On the face of it, that crucial destructive act implies that the spider has maintained throughout the process a thorough grasp of the overall design of the silken weapon she has built and is fully aware of the mechanics and limitations of the finished structure.

Rotating to a head-down position, the spider then carefully picks up the corners of the net with her four claw-tipped front feet and gathers it into loose horizontal folds. It is now time for a final check. Still holding the corners of the net, she extends each of her four front legs horizontally, one after the other, stretching the net diagonally in each direction to ensure that it is well made and fully operational. If it passes this test—I have never seen it fail—she then delicately reaches down to touch the ground with one leg to ensure that the drop is perfect and that her prey will pass within the range of her net-casting lunge. Finally, she settles down to wait.

Net-Caster Attack

Her livelihood now depends on her huge ruby hunting eyes, whose equivalent in photographic terms would be an fo.5 camera lens. They allow her to hunt even by starlight, and the slightest movement in the target zone beneath her will trigger her attack reflex. The four front legs will flick downward and outward, spreading the net to more than four times its original size. It catches on whatever it touches, and the spider then withdraws, lifting her prey off the ground, allowing it to thrash about in vain and become further entangled. When her victim is thoroughly enmeshed and immobilized, she wraps some fresh silk around it, injects a lethal dose of poison through long, curved fangs, and begins to feed. If she is hungry she may construct a new net while still feeding.

The net construction, taking some twenty minutes to complete, represents an engineering achievement of astonishing sophistication for an animal of any kind, let alone a species with so small a brain. It therefore poses an intriguing question for us humans: where does the spider store the memory and the decision-making machinery for such complex technology? While no one would suggest that a spider "thinks" as we do, the net-caster's nightly exhibit of technical and tactical proficiency must at least make us pause and reconsider the sources of our own behavior.

Tyrannies of Scale

The human brain contains around 100 billion neurons, and each of those neurons is connected to about 1,000 others, giving our brains a range of alternative synaptic connections that is, according to neurobiologist Richard Thompson, "larger than the total number of atomic particles that make up the known universe."[2] By contrast, a spider has only several thousand neurons to think with, and the total number of synaptic junctions linking them together is fewer than 1 million. But there is a downside to big brains. Our brains, and the almost 80 kilometers (45 miles) of neuronal circuitry attached to them, require large bodies and a disproportionate percentage of the body's overall energy just to keep them ticking over. The modern human brain represents a little more than 2% of the body's mass but requires a minimum of 16% of the blood supply. Spiders, on the other hand, are built for economy and simply do not have an energy budget large enough to run a disproportionately large brain.[3]

We might suspect that a spider's neurons would be much smaller than ours and that its brain is therefore comparatively large, but we would be wrong. A spider's neurons are virtually the same as ours, both in size and structure, and they use the same kind of neurotransmitter substances to regulate them. In other words, the capacity and flexibility of a spider's brain is directly comparable to a similar volume of human brain.[4] It is therefore a fairly safe bet that a spider can't think—in the human sense of the word. Yet, despite this limitation, spider behavior may still vary considerably between different individuals of the same species, particularly in response to physical threats. A common orb-web garden spider of the genus *Argiope*, for example, may use any one of three escape strategies when threatened by a human finger: it may immediately free-fall to the ground to hide in the grass or leaf litter, it may run along a lateral web-support to hide in the shrub or tree to which it is attached, or it may slip through the web so that it places the web between its body and the threatening finger. The choice they make seems to be wholly "personality-dependent." Three such spiders in my garden have each settled on a different response to the finger threat, and each uses their preferred response almost exclusively—even though all three strategies are equally available since the spiders are of the same species and build their webs close together between the same two shrubs.

A Chemistry in Common

Having spent much of his academic life studying the brains of arthropods, neurobiologist David Sandeman of the University of New South Wales finds nothing surprising in the net-caster's engineering talent or the garden spiders' behavioral flexibility. He admits to having the greatest admiration for spiders' highly specialized data-processing capacity and their ability to modify their behavior according to circumstances. Under the microscope, a cross section of a spider's brain displays well-defined internal structures that betray very sophisticated regional specialization—just as ours do. In fact, our neural pathways and neurochemistry seem to be common to most animals and probably evolved long before life emerged from the sea. Spiders even use such "human" neurotransmitters as serotonin, dopamine, and acetylcholine, and an inhibitor, gamma amino butyric acid, or GABA. These similarities have led Sandeman and many of his neurobiologist colleagues to conclude that "the way a spider's sensory input is integrated and eventually issued as motor commands appears to follow precisely the same rules as do our own nervous systems."[5]

Since young net caster spiders remain hidden near their birthplace for several days after they are born and then disperse to hunt independently, they never see their mother manufacturing her aerial blind or her complex weapon. We are therefore left with only one explanation for the similar behavior they all display soon after hatching: it must be encoded in their genes. Particular behavior sequences are then switched in or out according to whether circumstances match the mental templates embedded in its neuronal circuitry. But if this degree of behavioral complexity and capacity for fast, flexible "judgment" springs solely from spider DNA, then perhaps we should not too readily dismiss human DNA as the primary source of human behavior.

The same kind of complex genetic programming that retiarius spiders exhibit is visible to varying degrees in the behavior of all animals, but it is most unequivocally displayed by species in which the parents play no part whatever in nurturing their offspring. Species best known for this kind of complex genetic behavior include salmon, eels, marine turtles, and several migratory birds. For example, the female European cuckoo (*Cuculus canorus*) deserted by its own parents and raised by foster parents of another species, must find its own way to the cuckoo mating grounds in southern Africa each winter. On returning to Europe, the young female cuckoo must then instinctively adopt the devious egg-laying strategy of its unknown mother and surreptitiously deposit its eggs in the nests of suitable foster species. Here is genetic programming of astonishing complexity.

Genetic Reflexes

Even in those species whose lifestyles offer ample opportunity for parents to instruct offspring in appropriate patterns of behavior—in other words, species such as ourselves—the undercurrent of genetic imperatives is only too apparent. The evidence is most clearly displayed during infancy. After

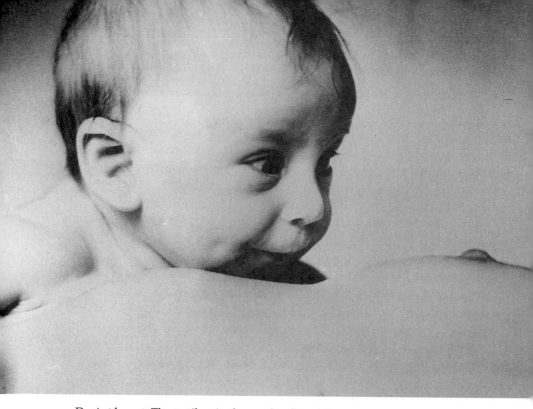

Dani at breast. The tactile, nipple-search reflex of the newborn is soon replaced by an experience-based, visually guided search exemplified here by my daughter Dani at about five months old.

that time, behavior is too easily attributed to parental and cultural influences. However, all newborn babies display reflex behaviors that cannot have been learned and must therefore be programmed into them. The baby's brain, like the spider's, performs like a computer terminal that is directly linked to the mainframe of its DNA. Stimulate the baby in the right way and, if the baby's neuronal system has been properly wired, it will display the standard human response. Touch the newborn on its cheek with a fingertip and it will turn toward the finger and begin to suck on it.

The baby's instinctive search for the nipple is a wholly genetic and obviously functional response. All mammals have it. But the human baby also comes equipped with dozens of other genetically programmed reflex behaviors that seem to have little or no functional link at that stage of development. However, most of these fade away during the first few months of development and never reappear. We can only surmise that they are the remains of genetic programming that must have worked for our evolutionary antecedents tens of millions of years ago when infants had to assume control of their bodies within a few minutes of birth. Among the best known of these short-lived behaviors are the hand-grasp reflex, the walking reflex, and the swimming reflex. They are displayed by all normal babies, both human and chimpanzee, in response to various stimuli. Even though some may be precursors to adult patterns of behavior—grasping, walking, and swimming—all have to be totally relearned at a later stage.[6]

One of the more complex and interesting of these genetically programmed behaviors, the Moro reflex, gives a clear indication of its original function. If a baby is supported horizontally, face upward, and the head is then allowed to fall backward, the infant's normal reaction is to arch its back and extend its legs. At the same time it reaches outward with both arms and then immediately pulls both inward toward each other. Chimp babies display precisely the same response, but for them it has very real value. Allied to the strong hand-grasp reflex, precisely such a sequence would provide the best possible insurance against falling from a hairy mother moving through the treetops.[7]

Similarly, a baby's temperament—irritability, shyness, passivity, and so on—also comes as part of the genetic package. Even the degree to which a baby will empathize with the distress of other babies seems to be a part of the inheritance. Some babies will almost always cry in response to the cries of other babies. Some rarely do. Identical twins, on the other hand, almost never cry solo, and if one cries in response to another baby's cries, then both twins will—a fact that signals the peculiar value of twins in the study of inherited behavior.

One Egg, or Two?

The paramount role played by genes in directing human behavior is as clearly demonstrated by the extreme similarity that invariably exists between identical, or monozygotic, twins, as by the differences between fraternal, or dizygotic, twins. A zygote is the cellular product of gametic union—a fertilized egg cell in the case of human beings. Monozygotic twins originate from a single fertilized egg cell; dizygotic twins originate from the coincidental fertilization of two separate egg cells.

As the product of a single fertilized egg that cloned itself before beginning to differentiate, monozygotic twins share the same genes and represent two expressions of a single animal. Fraternal twins, on the other hand, are no more closely related than are any siblings. Studies of twins consistently show vastly greater similarity in temperament between twins who share their genes than those who don't. Identical twins show correlations in the region of 90% for physical factors and 50% to 90% similarity for more subtle and disparate variables such as personality, brain-wave patterns, IQ, and the nature and timing of both major and minor decisions in their lives. Fraternal twins, like other normal siblings, never achieve such close correlations.[8]

For the past two decades Thomas Bouchard and a team of researchers from the University of Minnesota have tracked down and studied more than fifty pairs of identical twins who were separated at birth or soon afterward. Exhaustive analysis of the twins' personalities, behavior patterns, opinions, interests, intelligence, social values, and political and religious belief systems, as well as dietary and recreational habits, sex lives, and television viewing preferences, suggested that the degree to which life experiences can skew our inherited personality is small indeed. The data frequently showed that the twins had chosen similar career paths, hobbies, and interests and had acquired a similar spectrum of friends. Most showed strikingly similar patterns in their personal relationships, marrying partners with the same first names on the same day, having children of similar ages, and giving their children the same or similar names. Some twins were wholly unaware of each others' existence during these synchronous behavior patterns. Certainly having been raised by different adoptive parents in different locations and environments seems to have made very little difference to their overall belief systems and behavioral similarity. In fact, twins reared together by their natural parents often display more variation than twins reared apart. This is generally thought to be the result of the natural parents' determined efforts—and perhaps the twins' own efforts—to generate apparent difference where little or none existed at birth. It indicates that most of the minor differences between identical twins develop during the period of greatest brain plasticity, between birth and adolescence, when slightly different life experiences are able to produce minor differences in the brain circuitry as it develops. In other words, their genetic drives are identical, but subtle neuronal wiring differences may alter their expression.[9]

Genes and Attention-Deficit Hyperactivity Disorder

The childhood behavioral disorder commonly known as attention-deficit hyperactivity disorder (ADHD) is a common childhood problem, affecting between 4% and 6% of the school-age population. Involving restless, impulsive behavior and an inability to concentrate, ADHD is somewhat ill-defined and in its mild form appears to blend seamlessly into normal childhood behavior. In one of the largest studies yet undertaken, a research team from Australian and U.S. universities surveyed 2,000 Australian families, including 5,067 twins and siblings between ages four and twelve. When researchers matched the children's behavior against two of the most commonly accepted ADHD diagnostic scales, the results confirmed that ADHD did indeed vary on a continuous scale throughout the population, but it also showed quite clearly that the source of that behavior variation was primarily genetic rather than environmental. The rate of symptom sharing by monozygotic twins was approximately double that of fraternal twins, and the rate between siblings was less than half the dizygotic rate. Significantly, these rates of symptom sharing applied equally to boys and girls, so gender was not a factor. The percentage match between identical twins was most revealing. If one identical twin displayed the symptoms, then the chance that the other twin did also was somewhere between 75% and 91%. The overall figures suggested that the effects of a shared family environment contributed only about 13% to ADHD symptoms. DNA studies have shown that at least three genes are involved in the ADHD behavior.[10]

Such findings confirm two decades of similar work in England by Hans Eysenck, who in 1981 summed up his study of the genetic component in human behavior as follows: "The minimum value for the contribution of heredity is something like 50%, with the maximum being in the region of 80%. . . . Thus there can be no doubt that personality has a hard core of innately determined behavior patterns; this hard core (the 'genotype' as geneticists call it) interacts with the environment to produce the actually observed behavior (the 'phenotype')."[11] Although Eysenck's findings provoked a storm of dissent in the 1970s and early 1980s, the rising tide of genetic and neurological evidence in his favor now makes his claim seem moderate indeed. Many geneticists, including Bouchard, now believe that no dimension of our behavior is wholly immune to the effects of genetic expression." In fact, I would suggest that the moment we accept that we are entirely normal animals, then no other reasonable conclusion is available to us. The differences and similarities in the behavior patterns of identical and fraternal twins point unequivocally to the genetic origins of all human behavior and the very practical unity that exists between mind and body in all animals.

To See Is to Perceive

A neurological illustration of this mind-body unity occurs in the process of seeing. Until the 1970s it was believed that the act of seeing was a two-stage process. It was thought that images of the outside world became electrically imprinted on the retina of the eye like digitized maps that were immediately relayed by the optic nerves to a kind of blank screen in the occipital cortex at the back of the brain. The informational "picture" registered on this screen was then supposedly analyzed by "the abstract mind" and, after due reference to other remembered images, assigned various meanings.

Nothing of the sort occurs. To quote Semir Zeki, professor of neurobiology at the University of London, there is, indeed, a "profound division of labour within the visual cortex," not between the physical and the "spiritual" but between four parallel physical systems that simultaneously register different attributes of vision.[12] One neuronal system registers motion, one registers color, and the other two specialize in different aspects of form. Of the two form-sensitive systems, one is intimately linked to color and the other is independent of it. Although these systems are clearly distinct, their anatomy nevertheless allows all four to register and analyze the image at the same time. A vast complex of neuronal interconnections allows them plenty of opportunities to communicate with one another and with other regions of the brain *during* the process of data transmission.[13] The considerable advantages offered by exactly this kind of parallel data processing are currently being exploited in the computer industry with spectacular results.

The multistranded nature of the seeing process is confirmed by patients with lesions in the visual cortex that are so localized that they inhibit the operation of only one of its four specialized systems. Some patients see only in shades of gray. Some can see only stationary objects—should an object move, it disappears from view. And some patients with damage to the linkages between the undamaged seeing systems can see and even draw what they see, but they can make no sense of the image, either in the brain or in their drawing, even though they might be entirely rational in other ways. In these patients the normal cross-referencing fails to take place during the seeing process.

According to Zeki, "It is no longer possible to divide the process of seeing from that of understanding, as neurobiologists once imagined, nor is it possible to separate the acquisition of visual knowledge from consciousness. Indeed consciousness is the property of the complex neural apparatus that the brain developed to acquire knowledge."[14] The logical inference, then, is that there is no clear-cut distinction between mind and body, but rather that they function as a single entity.

A Programmed "Error"

Most of us still find it not only distasteful but practically impossible to surrender our belief in the existence of a separate and independent "mind" or "spirit." So despite mounting evidence for the biological unity of mind and body, the proposition will never be a serious threat to the everyday world that lies beyond the laboratory door. The universality of belief in the autonomy of the human spirit—the mind-body duality—is a powerful indication that this belief originates in our genetic makeup. Without that crucial, genetically generated drive to mysticize our perceptions, culture could not have become such a powerful feedback mechanism, reinforcing as it does the genetic behavior of successive generations by continually reimposing such altruistic ideals as duty, honor, patriotism, and "family values." Because culture constitutes such an effective genetic feedback mechanism, we must conclude that the morality we so diligently pursue in the name of personal or tribal integrity is no more than a shrewdly fashioned genetic propaganda device, a device specifically designed to heighten our mystical gullibility and conceal from us the real source of our behavior—our genes.

In linking the words *genetic* and *behavior,* I do not mean to suggest that particular genes code for particular behaviors. In fact, it is now generally conceded that many genes are multifunctional, and large numbers of them undoubtedly play minor roles in most behavior. Of the 100,000 genes we inherit from our parents, it is believed that about half are directly responsible for the developmental wiring and maintenance of the brain, and even these are not the final arbiters of its circuitry, which normally involves quadrillions of synaptic junctions—far too many to be specified by a mere 50,000 genes. At birth, a baby's brain contains all the neurons it will ever have—about 100 billion. But its final wiring pattern will depend on which circuits are used most during the next eight to ten years of its life. After that time, those synapses that have been rarely used undergo a dramatic culling process, which in extreme cases of sensory deprivation may result in a brain that is up to 30% smaller than is typical.

Like a computer, the human brain is wholly dependent on the quantity and quality of the data fed into it. It can only store and interpret that data in the genetic "language" system it inherited in its DNA, and it can respond only in the behavioral vocabulary embedded in its genome, just as all animal brains do. We may agonize over alternative courses of action for as long as we like, using whatever combination of reason and intuition we feel is appropriate, but our final decisions still represent the inevitable reactions of our particular genetic makeup to the peculiar patterns of perceived information investing us at the time. In other words, the choices we make are those we cannot help making in the circumstances.

Our Indomitable DNA

To those who recoil from such concepts, I would suggest that a genetic explanation for human behavior is vastly less deterministic than the alternative proposition—that our environment, social or natural, has a controlling influence on us. The uniqueness of each human genotype ensures that although we may be subjected to the same environmental pressures as our neighbor, unless he or she happens to be a monozygotic sibling of ours, there is absolutely no possibility that we will both react in precisely the same way. In other words, it is our genes that ultimately underwrite our freedom from the tyranny of external control. In fact, the concept that all our behavior is genetic in origin should displease only those who wish to manipulate us. The uniqueness of our dictatorial genes guarantees our unpredictability and the ultimate failure of any who attempt to control our behavior. Thanks to the autonomy of our DNA, we remain indomitable and therefore free in the only sense that really matters to us.

It has been argued that something called "genetic determinism" is a suspect concept in that it offers too easy an excuse for past actions ("I couldn't help it, Your Honor, my genes made me do it") and a convenient means of avoiding responsibility for future ones ("I have no choice but to do it my way"). This argument fails to account for the fact that we could only take cynical advantage of those excuses if the central proposition were *untrue*. As it is, however, even that degree of free will remains unavailable to us, and for very good reason. If we were not continually at the beck and call of our genetic code, and if we did not instinctively believe in the duality of existence and act accordingly, then our mythically driven civilization would grind to a halt. In other words, our delusions of spiritual autonomy are essential to the survival of our species.

Finally, a proper understanding of our genetic subservience would also rob us of all our heroes and villains, since they too are governed by their genetic imperatives and are no more worthy of reverence or condemnation than the rest of us. "Thought, however unintelligible it may be, seems as much function of organ, as bile of liver," wrote Charles Darwin. "This view should teach one profound humility, one deserves no credit for anything. [N]or ought one to blame others." Unfortunately, this concept is as unacceptable now as it was when Darwin first penned it in one of his notebooks in the 1870s. (He wisely refrained from pursuing the point in his published works.)

Unwitting Darwinians

Although we are generally secure in our delusions, there is a small army of unwitting Darwinians who bet on a daily basis that he was absolutely right. In fact, they regularly stake their professional reputations on the proposition that every decision we make is the product, not of reason, but of our very own genetic code operating in secret within each cell of our bodies. These true believers are the world's advertising agents, and most of them make money on the bet.

Although few seem to comprehend the genetic nature of their wager, by carefully targeting our overriding imperatives for sex, procreation, security, tribal status, and some form of mystical belief system, advertising agencies find that they can unload almost anything, a packet of popcorn or a presidential candidate. No matter how irrelevant, dangerous, or impractical the product, if it can be convincingly attached to one of our basic genetic imperatives, sales are virtually guaranteed. In fact, the only effective blocking device is another genetic imperative operating competitively within the buyer. The deal might appear, both to the buyer and the seller, to be negotiated in the language of reason and logic, but this is window dressing, designed to disguise—especially from the buyer—the genetically determined feelings that are at work here. A flashy car, trendy house, or expensive stereo system might represent increased tribal status (and therefore sexual advantage) to the genes of a young male buyer, but these attractions must be weighed against the threat to future status and security represented by a hemorrhaging bank account. Which option he ultimately takes depends largely on the relative strengths of his contending genetic imperatives. This example is simplistic, but similar kinds of concealed genetic negotiations underpin all commercial and social transactions. The daily successes and failures of the advertising world provide, not a yardstick of human materialism and gullibility, but an accurate measure of our genetic reactions to a rapidly changing social and physical environment. If we accept that we are typical products of evolution and are, like all animals, genetically driven in all behavior, then no other explanation is available to us. And the key to that behavior is emotion.

Raising the Gene Flag

We can easily tell the moment our genes cut in because it is then that our emotions become aroused. The presence of emotion is a clear signal that one or more of our genetic imperatives have been triggered and the old mammalian-reptilian structures of the brain have taken control. And whenever that old fraud "morality" strides onto the stage—usually dressed in the noble regalia of Duty, Honor, Loyalty, Patriotism, or Justice—it signals that our triggered imperatives are in open conflict and gangs of genes are at war with each other. If more information becomes available, the outcome may change, but only if the new information ups the ante for the contending set of genes. Whether such a change of attitude occurs is governed by the genetic value implicit in the incoming information and by our reassessment of the relative risks attached to the behavioral alternatives. The only trouble is that once primary imperatives like sex, tribal status, or territoriality have been aroused, it becomes very difficult indeed to smuggle new factual information into the argument.

Human beings are only one of millions of species that have to deal with these kinds of internal genetic conflict and make difficult, and often expensive, compromise arrangements. Those species that are confined to sexual reproduction face continual conflict of this kind, since all biological devices capable of buying genetic advantage are inherently risky and therefore genetically controversial. The price is always high, whether it be to buy a fast car, grow a fine set of antlers, or even produce the smallest flower. Make the wrong evolutionary decision and it may be curtains for your particular genetic heritage.

Risky Advertisements

Flowers, for example, are a valuable aid to pollination, and therefore to propagation and genetic survival, but they are both expensive and risky for the plant that grows them. Not only must the plant set aside valuable energy and materials to manufacture nutritionally unproductive accessories like petals, coloring matter, perfume, and nectar, but also such gaudy advertisements may attract herbivores as well as pollinators. Similarly, the spectacularly overdeveloped plumage of birds such as peacocks, birds-of-paradise, and argus pheasants might be irresistibly seductive to the female of the species, but it also represents considerable expense and serious handicap to the males that wear it. "The long tail of the peacock and the long tail and wing-feathers of the argus pheasant must render them an easier prey to any prowling tiger-cat than would otherwise be the case," wrote Darwin. "Nevertheless I know of no fact in natural history more wonderful than that the female argus pheasant should appreciate the exquisite shading of the ball-and-socket ornaments and the elegant patterns on the wing-feathers of the male."[15] In general, however, devices such as flowers and fancy feathers are relatively truthful advertisements of genetic fitness, and the goods they advertise are in fact worth the risk or there would be no flowering plants and no peacocks as we know them.

It is the same for the young car buyer. Urged on by his emotions, he is tempted to surrender to a genetic advertisement that is inherently risky and expensive in every sense. In closing the deal, his genes have to be convinced that they are about to gain more than they will lose. Since financial ruin tends to weed out those status-seeking males who habitually misjudge the risks and cannot really afford the outlay in the first place, the ownership of an expensive sports car is a relatively honest advertisement of genetic fitness in an evolutionary sense. In other words, the owner not

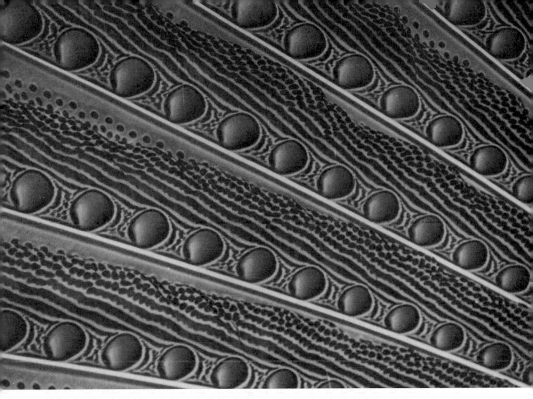

Tail feathers of the great argus pheasant (*Argusianus argus*). The male argus pheasant attracts a mate by displaying his elaborately shaded tail and wing feathers. Darwin was taken with the female argus pheasant's obvious appreciation of the elegant display.

only projects success in terms of tribal status but also appears to be financially able to provide food and shelter for any offspring that might arise from a sexual union. Herein lies the answer to the ancient riddle: why do so many good women get involved with such "bastards"? Women find them attractive because female genes demand that they take the risk of mating with well-heeled, ambitious, warrior males so that they might be better represented in the next generation. By marrying for tribal status in this way, the woman is gambling that her children will not only have some degree of tribally backed security during their vulnerable years but that her genes will be partnered by strong male genes in their combined assault on the future.

Our Short-Sighted Genes

Sometimes however, the woman's genes are so preoccupied with achieving these immediate advantages for her descendants that they fail to recognize signs of genetic imbalance concealed in the flashy warrior package.

Such would-be warriors may be so obsessed with attaining tribal power and status, and so frustrated by their inability to achieve them, that they are reduced to beating their women to a pulp to satisfy their lopsided imperatives. Genetic blindness to these long-term risks routinely costs many women their lives.

Anti-smoking campaigners face similar genetic blindness when they try to persuade adolescent genes to take note of the health warnings on cigarette packs. Unfortunately for adolescents, the vague prospect of impaired health in the distant future offers no competition whatever to the immediate prospect of being seen as an experienced, free-thinking female, or a brave, indomitable male. This is the stuff tribal legends are made of. And if it pays to advertise, it pays even better to exaggerate. Whether you are a politician or a puffer fish, deceit is a necessary part of survival. If you can fool a potential aggressor into thinking you are more dangerous or distasteful than you really are, then you may live another day—and reproduce yet one more time. If you can make yourself more attractive by some artifice or other, then this too may multiply your mating options.

Just as some species of plants have evolved flowers that are larger and more decorative, or better stocked with perfume and nectar than mere functionality demands, so our species seems entirely unable to resist the urge to gild the lily with the aid of tattoos, jewelry, feathers, perfume, artificial hair, clothes, and cosmetics. Yet if such ploys did not work at the genetic level, they would not persist within the species. Consequently, the human imperatives to survive and reproduce have given rise to far more complex genetic advertisements than flowers and feathers. Aided by technological ingenuity and a catastrophic population explosion, those imperatives have led to a new kind of global disease—consumerism. This disease has already begun to tear the biota of this planet into tatters. Our constant attempts to express ourselves and to be successful—in other words, to advertise our genetic worth—now entail massive consumption of earth's energy and resources. Every scrap of solid, liquid, or gas that we produce in our pursuit of those genetic imperatives, thereby launching our genes into the future more successfully, is part of our natural biological waste, in the truest sense.

Inevitably, most of this extravagant consumption and pollution is a byproduct of the genetic code of the male rather than the female. The genetic fitness of the human male has always been publicly defined by his tribal status and efficiency as a provider of food and security for his family. That fitness used to be advertised merely by a strong, athletic body, ornamented by symbolic scars, tribal tattoos, painted designs, feathers, penis decorations, combat trophies, or other generally accepted badges of tribal worth. But when farming and commerce replaced hunting and gathering, those old advertisements became inadequate and irrelevant. It was essential that males evolve new symbols and new displays to advertise their genetic superiority in this new and very different setting. The warrior was about to transform himself into the human equivalent of an Australian bowerbird.

The Bower Builders

To attract a mate, male bowerbirds construct elaborately decorated stages out of twigs, leaves, and mosses and devote inordinate attention to their maintenance. The satin bowerbird (*Ptilinorhynchus violaceus*), as a compulsive collector of blue baubles, frequently decorates his bower with blue plastic clothespins. The arrival of a curious female prompts the male to begin an elaborate courtship dance, which if successful concludes with a brief, noisy mating. When they part, she flies off to single parenthood while the male re-baits his love trap and waits for the next female. So important are these theatrical settings to the male bird in his struggle to achieve sexual fulfillment and genetic success that in some species neighboring males undertake lightning raids into each others' territory to tear down their rival's stage and steal his decorative baubles.

Propelled by similar genetic imperatives, human males have, during the past 10,000 years of diligent bower building, dramatically redesigned the currency of their genetic fitness in a similar fashion. The most successful performers now tend to leave behind an astonishing assortment of concrete, steel, bricks, mortar, heavy metals, plastics, pesticides, and other toxic wastes as the normal detritus of their bowerbird existence. These substances are just as much a part of the biological waste of *Homo sapiens* as the discarded tail feather of a peacock or the twiggy bower and blue clothespins of the satin bowerbird. Similarly, the annual turnover of the world's advertising industry is not so much a measure of modern materialism as a precise flowchart of humanity's culturally modulated genetic imperatives.

We simply do not possess the foresight to perceive the point at which the long-term risks inherent in our astonishingly expensive genetic advertisements outweigh the advantages. As with cigarettes, the health warning on our glitzy cultural package is too ill-defined and too far removed to affect our here-and-now biological priorities. And besides, we have been specially fitted with a very effective blindfold.

Practically Deranged

During humanity's developmental period, its DNA was busy nurturing far more important behavioral characteristics in our hominid ancestors

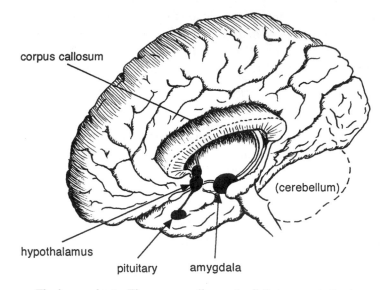

corpus callosum

(cerebellum)

hypothalamus

pituitary   amygdala

Figure 21. The human brain. The corpus callosum is all that connects the two cortical hemispheres of the brain.

than an aesthetic preference for factual explanations of natural events. As hominid brains began to bulge with new neuronal networks, and as their capacity to reason increased, they might well have been in danger of dulling the fast reactions and savage responses inherited from their very successful primate ancestors. Reasoned thought had to be kept separate and subordinate to the tried and true system of nonrational instinctive reactions. If they had attempted to work out and weigh up the long-term evolutionary values of alternative behaviors whenever instinctive re-actions were required, the time delay would, in most instances, have ensured that they not only lost all advantageous moments but their lives as well. But so long as their capacity for rational thought remained an optional extra, the solution was both simple and practical: whenever mat-ters affecting genetic survival arose, the genes assumed direct control and rational government gave way to martial law.

Paradoxically, such automatic abdication of rational thought would have been especially valuable in the face of imminent destruction. If our ancestors had all turned and run whenever reason dictated that they should, then *Homo* DNA would have come to its very logical end in a few patches of blood-stained African dust some 2 million years ago. As prey animals, humans were even more ill-equipped to run than they were to turn and fight. If evolution had added the leaden handicap of logic on such occasions, it would have guaranteed the leopards a regular lunch of tender young humans, and none would have made it to old age. There is little doubt that during the past 2 million years of human evolution the cold-blooded processes of Darwinian selection would have unerringly weeded out many a deep thinker in favor of the wild-eyed fanatic—among tribal leaders especially. The clinical eye and the cool head would still have had their uses, of course, but only as optional extras that could be called on to solve tactical or technical problems and then be relegated to their usual subordinate role—just as they are today. But with mystics leading the tribe, evolution faced a new and knotty problem. How do you combine spectacularly irrational behavior with a gradually increasing capacity for rational thought and still keep the brain running relatively smoothly?

The solution seems to have been provided by the natural physical division of the brain into two separate hemispheres. Almost all of the information that passes between the brain's two hemispheres travels via the corpus callosum (see figure 21). Evolution seems to have capitalized on this by making the two hemispheres responsible for very different functions and then limiting the interplay between the two sides—a classic case of divide-and-rule.

Our Split Brain

After a long series of experiments with patients whose brain hemispheres had been surgically separated (by cutting the straplike corpus callosum that directly links them), neurobiologist Roger Sperry found himself forced to conclude that "surgery has left these people with two separate minds, that is, two separate spheres of consciousness." He added that "this mental dimension has been demonstrated in regard to perception, cognition, volition, learning and memory."[16] In most cases, severing the corpus callosum separated the right hemisphere from its only means of communication with the outside world, the left hemisphere's speech factory, otherwise known as Broca's area. In one extraordinary case, however, a split-brain patient who had sustained some left-hemisphere brain damage as a child revealed verbal competence in both hemispheres after surgery. Sperry and his colleagues were then able to communicate with each hemisphere separately and discovered during extensive tests designed to reveal the patient's personality that two entirely separate characters inhabited the two hemispheres.

Equipped in this fashion with the capacity to operate on two levels of awareness while being conscious of only one, our hominid ancestors were sitting ducks for the evolutionary sting that followed. It is hard to imagine a shrewder mental "flaw," even if evolution had intended it. That shadowy gap between the two spheres of human awareness left genes with precisely the loophole they needed to retain ultimate control of the body's entire communications system. If the analytical and constructive hemisphere, the left, was not at all times fully aware of the wide range of perceptive activity occurring in the right hemisphere, then here was a gap in the cortical defenses through which whole truckloads of mystical nonsense might pass virtually unchallenged.

The significance of this gap in our cortical defenses was brought into sharp focus for the first time during experiments conducted by Michael S. Gazzaniga and his colleagues at the Center for Cognitive Neuroscience, Dartmouth College. The experiments hinged on the fact that in most people only the left hemisphere of the brain has a "voice"—provided by Broca's language area. The highly perceptive right hemisphere must therefore relay all of its findings via the corpus callosum to this specialized language center before the data can be translated into meaningful communication. This also means that all right-brain messages may be subject to left-brain censorship, and perhaps considerable creative editing, before they are cleared for publication.

In one seminal experiment, a split-brain subject was presented with two large pictures: in front of the left eye, an urban snow scene, and in front of the right, an image of a bird's foot. Below each of these pictures were four smaller images, only one of which could be related in any reasonable way to the large picture above it. When asked to point out an image relationship with the right hand, the subject correctly linked the bird's foot to a picture of a chicken. Asked to make a similar relationship choice with the left hand, the subject's left index finger correctly picked out a picture of a shovel as having the most relevance to the snow scene above it. The subject was then asked to explain the decisions. The verbally accomplished left hemisphere immediately offered the obvious explanation for its own decision to link the chicken with the bird's foot. When asked why the left hand (right brain) had chosen the shovel, the subject's surgically isolated left brain should have admitted that it didn't know. But to the researchers' surprise, the subject replied that the shovel had been selected because it would be useful for cleaning out the chicken shed. Clearly, the left brain had fabricated this memory in order to conceal its total ignorance of its counterpart's activity.

It seems that our loquacious left brain cannot abide a vacuum. As it ghostwrites our right-brain narrative, it obsessively fills in any gaps and injects snippets of its own propaganda wherever it can. Here, then, is the source of the so-called false-memory syndrome, and no doubt the origin of most of our mystic visions and spiritual fantasies.[17] According to Sally Springer and Georg Deutsch, "The right hemisphere has a suspicious emotional tone, for it sees conspiracies where they don't exist as well as where they do. It needs the left hemisphere to analyze critically the patterns it generates in order to test their reality."[18] Significantly, the one area of the brain where sexual dimorphism is most evident is the corpus callosum. The female version is thicker and more bulbous than the male version and has many more nerve fibers linking the two hemispheres. Consequently, men have much poorer communication between the left and right sides of their brain.[19] It is as if evolution specifically widened the brain gap in men to ease the passage of their elaborate fantasies.

Our Vital Insanity

To properly accommodate this vital streak of insanity in an increasingly rational brain, it was first necessary for people to perceive, quite accurately, that their genetic imperatives—their instincts, feelings, morality, and religious beliefs—represented a source of considerable wisdom, even supernatural power. Added to this was the less accurate belief that this inner source had its roots in an invisible world of superintelligence, a mystical world that lay beyond rational comprehension.

Here evolution had hit on the sweetest of solutions. Such perceptions were guaranteed to produce a faith-dependent species that believed itself to be thoroughly separate from the rest of the animal kingdom but followed its genetic instructions to the letter—and left more offspring as a consequence. Here was a gene-driven animal just like any other, yet one that believed itself to be under special guidance—guidance that was not merely spiritual, but in most instances divine. Here was a wonderfully practical insanity, an invincible hereditary madness that eventually enabled this underendowed paragon of animals to devour the planet like a ripe fruit.

Each vertebrate species has its own suite of mental derangements. Some provide irrational aggression when home territory is threatened by an invader, others enable the performance of bizarre courtship and other social rituals that have evolved within the species. Some of these nonrational behaviors may even add significantly to the performers' risk, the mating display of the peacock being an obvious example. Nevertheless, if the long-term dividend to the gene pool is greater than the immediate cost, then the behavior is likely to remain securely embedded in the population. Primates exhibit some of these nonrational but vital behaviors, although none are particularly bizarre or risky. In humans, however, these derangements have been switched into overdrive. This probably occurred in order to minimize the immense risks inherent in the major brain enlargement that began about 3 million years ago.

The human brain has doubled its volume and quadrupled the surface area of its rational cortex in that time, a degree of enlargement unprecedented in the evolution of any other species.[20] If behavioral control had gradually transferred from the instincts to the rational brain during this period—as is commonly assumed—I believe our end would have been bloody and swift. Even today, given our tenuous grasp of evolution and its complexities, the most genetically advantageous behavior usually lies far beyond the scope of instant rational computation. A million years ago too much rational thought would have been suicidal. In other words, without a genetic override mechanism securely wired into the brain of *Homo erectus*, that cortical enlargement would, I believe, have been lethal. Armed with an X-factor, an automatic override device that cuts off rational thought at a moment's notice and draws directly from a reservoir of pretested genetic behavior, we remained fully functional animals. This device enabled us to continue to feed, mate, and reproduce without interference from our enlarged cortex. To put it yet another way, our neuronal circuitry remained hot-wired to our genes so that we would not be handicapped by logic when genetic responses were called for. That is why, under the spell of our carefully programmed spirituality, we cannot help falling in love, yearning for sexual gratification, nurturing our children, forging tribal bonds, suspecting strangers, uniting against common enemies, and on occasion laying down our lives for family, friends, or tribe. No gene could ask for more.

"Significance Junkies"

We habitually attach some degree of mystical significance to anything that has a bearing on the survival of our genes, now or in the future, and this extends to the very edge of our perceptions and the limits of rationality. Some of us even give names and ascribe personalities to our cars and boats, touch wood for luck, or feel that things go better for us if we wear particular items of clothing. The late Carl Sagan expressed it perfectly: "We're significance junkies."[21]

The brain structures and neuronal pathways that achieve these crucial responses have now been generally identified. They lie, as might be expected, in the older regions at the core of the brain, and the unmistakable sign that they have been switched on is the presence of emotion. The brain structure primarily involved in the arousal of emotions is the hypothalamus, which in conjunction with the hormone-secreting pituitary gland that sits just below it seems to act as the control center for a wide variety of phenomena, including structural growth and the physical expression of mental states. Electrical stimulation of one part of the hypo-

Autistic/hypothalamus attack. Where the usual checks and balances of the frontal cortex have been inhibited in some fashion, such as in those suffering from the brain dysfunction known as autism, the amygdala and hypothalamus retain full control, leaving the sufferer prone to outbursts of incoherent rage and frustration.

thalamus can trigger rage and a full-blown attack response in humans and other mammals; stimulation of a neighboring part of the hypothalamus can elicit feelings of intense pleasure.[22]

The hypothalamus is also linked by an intricate web of neurons to all the other ancient mammalian-reptilian structures that form the core of the brain. Sitting at the hub of this vast network of nerve fibers is the almond-sized amygdala. Functioning like a telephone exchange, it provides fast, reciprocal links between the brain's older components and the body's autonomic nervous system. The amygdala seems to be primarily involved in appraising the genetic significance of situations in which an immediate threat might be involved. In other words, it governs the primordial fight-or-flight reflex and determines the particular thresholds of aggression and discretion that characterize each of us. It also seems to play a major role in determining the nature and intensity of the emotional memories that help to define us as animals. These links put the amygdala, the hypothalamus, and indeed all the other ancient brain structures, in direct control of the hormone secreting endocrine system, thereby providing our genes with a kind of automatic choke that is able to flood the entire body with the appropriate chemistry to take us from standing start to top gear without a moment's thought.[23]

Logic, the Optional Extra

To summarize, not only are there two spheres of awareness available to us, but two entirely separate behavior control mechanisms, one rational and one entirely nonrational. Judging by the relatively recent enlargement of the cerebral cortex, and the fact that all the primary body controls have nevertheless remained securely in the grip of the much older nonrational system, the rational brain should be viewed, not as the principal generator of behavior and the pivot on which the species turns, but as an optional extra designed to be switched off the moment any serious evolutionary matters, such as genetic survival or propagation, arise.

For this evolutionarily useful characteristic to persist in our peculiarly rational species, however, it is absolutely essential that we continue to mistake its genetic origin for a supernatural one and, like the French dramatist Voltaire, continue to have faith in belief when belief is beyond reason. Even the waspish Voltaire never realized quite how close to the genetic truth he came when he wrote, "If God did not already exist it would be necessary to invent him." Precisely what we believe is immaterial; what matters is the kind of behavior it generates. This is why humanity is characterized by such astonishing diversity in its belief systems. As far as our genes are concerned, we can believe that the universe is driven by an overweight fairy on a green cheese bicycle provided that such belief effectively coerces us into adopting genetically advantageous behavior in all matters of evolutionary consequence, such as feeding, mating, nurturing, bonding, and protecting family, tribe, and territory.

Fitted for Faith

A curious link between religious belief and the amygdala surfaced in 1997 after a team of Californian scientists noticed that people with a brain dysfunction known as temporal lobe epilepsy (TLE) were often obsessively religious. Since repeated seizures within the temporal lobe are known to strengthen neural connections between the inferior temporal cortex and the amygdala, the researchers wondered if it might have a bearing on the level of religious fervor their TLE sufferers displayed. As the midwife of emotions and feelings of significance, religious or otherwise, the amygdala seemed a likely source of such passion. To explore this possibility, the researchers conducted a series of verbal tests on a group of volunteers, some of them TLE patients, some avowedly religious people, and some with no known religious commitment. The volunteers were hooked up to equipment that would measure changes in the conductance of the skin of their left hands, thereby measuring emotional arousal and heightened communication between the inferior temporal lobe and the amygdala. A list of forty unrelated words with various associations was then read to them by one of the research team. Some words were related to violence, some to sex, some to religion, and some were neutral. When the researchers matched the conductance readouts to the taped record of the word list, they found that only sexual words had given nonreligious subjects sweaty palms, whereas the religious subjects had been aroused by both sexual and religious words. The TLE patients, however, had become disproportionately aroused by religious words only.[24] While the very distinctive selective arousal in the three groups is interesting in itself, the TLE results appear to confirm the proposition that even our loftiest feelings of spirituality have their origins deep in the mammalian-reptilian core of the brain, a region that lies far beyond the reach of the rational cortex, a region ruled entirely by the genes.

The Cultural Constant

Whenever instinctive beliefs and emotional feelings arise, we may be sure that genetic imperatives have assumed control and our rational cortex has been bypassed. There is no reason to suppose it was any different 10,000 years ago, or even 1 million years ago. In fact, the deeper we probe into the past, the clearer it becomes that the massive complex of customs, rituals, faiths, and social mores that help to regulate our lives under the banner of culture are simply relics of previously successful genetic behavior—behavior that contributed so significantly to human survival that it became embedded in the society's cultural foundations. In other words, the iconography, symbolism, art, ritual, and other embellishments of a culture might alter continually and dramatically with the passage of time, but the underlying theme and central purpose remains invariable: to reinforce the pair-bond and tribal bonds, to subordinate and coordinate the tribal group, and to inspire altruism and in some instances extreme aggression in order to achieve genetic survival.

So, although our species' conquest of the planet might appear to represent the gradual triumph of the intellect over our brutish nature, in fact precisely the reverse is true. Being primarily founded on and driven by mystical beliefs of one kind or another, human civilization represents not so much a triumph of the mind over the body as the triumph of the gene over gene-threatening rational thought.

Let's take some simple examples from the arena of tribal behavior. One of the clearest illustrations of genetically manufactured cultural feedback can be seen in the phenomenon of teenage peer pressure. By resurrecting tribal regimes of loyalty, discipline, and territoriality and reimposing them on groups of teenagers, peer pressure ensures that our ancient tribal imperatives consistently override rational thought during those crucial fertile years. Similarly, if a young mafioso, a Triad warrior, or a member of a biker gang fails to put tribal loyalty above all else, then his adoptive tribe swiftly reeducates or ejects him. Likewise, the young executive who fails to bond with the corporate tribe and work ridiculously long hours at the expense of his marriage and family often finds that his career is either stunted or short.

Polygyny Pays

Wealth, tribal status, and political power are the social carrots that male genes constantly dangle in front of their gullible hosts in an effort to induce behavior that might achieve a wider representation of those genes in the next generation. The warrior who takes the bait, proves his worth, and achieves high tribal rank is then able to broadcast his genes more widely and more securely than his competitors. In other words, sex and political power are genetically linked, and leadership and promiscuity tend to go hand in hand. In fact, wherever a culture's male elite becomes especially dominant, it tends to institute either some form of legal polygyny or a social system that allows its members to have a multitude of morganatic mistresses as well as a wife or two. Less troubled by cultural restraints, despots usually opt for the simpler expedient of the harem, and some rulers are known to have mated with hundreds of women—and sired hundreds of children. According to anthropologist Laura Betzig of the University of Michigan, Ann Arbor, the record for polygyny seems to have been set about 2,500 years ago by an Indian potentate named Udayama, who kept a harem of 16,000 wives and concubines. As Betzig points out, however, these figures should not surprise us, since the polygyny principle is universal and governed by only three factors: the size of the population, the degree of political inequality in the society, and the disparity of wealth between richest and poorest. In small, simple tribal societies, the strongest men usually kept up to ten women for themselves; in more complex, medium-sized societies like the Samoans and other Polynesian cultures, the top men maintained access to as many as a hundred women; and in the largest societies, from the earliest civilizations onward, powerful leaders usually kept hundreds, thousands, even tens of thousands of women for their personal use. Lesser men in such societies also tended to be polygynous, keeping progressively fewer women according to their declining scale of status and wealth.[25]

Although the source of this extreme form of male chauvinism is customarily assumed to be an obsession with sexuality, power, and tribal status, in this behavior too the man is merely the puppet of his reproductive genes. The clue lies in the genetic payoff. The largest recorded dividend is 888 children, all sired by one despot, the Emperor Moulay Ismail the Bloodthirsty of Morocco.[26] Genetically speaking, polygyny pays. Humans are therefore inherently polygynous and always have been, and the politics of wealth and power are, in the final analysis, all about sex and genetic propagation.

Chastity's Other Face

Conversely, powerful leaders who show little or no sign of sexual promiscuity—such as Adolph Hitler—may be controlled by genetic imperatives so out of balance that they represent a far greater threat to humanity in the long term. Abraham Lincoln's shrewd aphorism that one should be wary of leaders with few vices has impeccable genetic credentials, especially where sex is concerned. In other words, a powerful male who *fails* to seek opportunities to convert his status into a genetic dividend is demonstrating that there is an error in the primary drive of his genome. With a deformity of such significance lurking in the unstable parliament of his genes, the potential for extreme behavior increases dramatically. In Hitler's case, it resulted in megalomania and genocide. Consequently, national leaders who are discreetly promiscuous are merely displaying reassuring evidence of their well-balanced ambition and general genetic fitness for leadership.

The male homosexual, on the other hand, appears (to male genes) to represent a very real long-term threat in that he will probably contribute little to the safety or fertility of the tribe and appears to represent a dangerous defect in the gene pool. This perception is the source of homophobia. The hysterical aggression that usually accompanies homophobic behavior makes it very plain that the reaction is genetic and reason plays no part here. Passions rule so that the macho genes can quietly get on with the assassination without interference from the rational cortex.

By such devious means, human genes have always used culture like a long-range whip to keep the dangerously rational beast on the straight and narrow path of genetically advantageous behavior. And it is not only males who are under the whip. That some form of marriage ritual, promoted and maintained primarily by women, is common to all cultures clearly signals its genetic origins. Its evolutionary role is unequivocal: to reinforce the pair-bond and to give any children of the union the safest possible start in life. By this stratagem, human DNA is able to apply a powerful external clamp to the relatively new and fragile phenomenon of long-term pairing. Such bizarre bonding behavior is rare in the primate world and does not appear in any of the other great apes.[27]

Long-term pairing would have become genetically advantageous for humans only after their maturation rate slowed down some 2 million years ago and children became such a long-term burden. Once that happened, the development of a culturally sanctioned system of pair-bonding would have added significantly to the survival prospects of its genetic product. If the man became a little more attentive at home, helped with the kids' education, and cooperated more closely with the other warriors in the daily business of hunting game and keeping the local leopards, hyenas, and human intruders at bay, then his own slow-maturing offspring stood a far better chance of surviving to sexual maturity. In other words, his genetically derived altruism provided the glue that not only bound the biological family together to form the basic social unit but also helped to weld the tribe's warrior group into an efficient hunting pack and a formidable fighting force.

The Antidote to Reason

To summarize, the ultimate origin of all human behavior resides in our genes, just as it does for all earthly organisms. It is vital, for sound evolutionary reasons, that we not see it this way, however. So in order to maintain nonrational, genetically determined behavior in all matters of evolutionary consequence, our genes are forced to strut their stuff behind a mask of emotion and culture. In this sense, it could be said that our mystically based cultures evolved specifically to counter the massive expansion of the human cortex and offer our genes the perfect antidote to critical analysis and reasoned thought in gene-threatening circumstances. If the human facility for rational thought had appeared 2 million years ago without the provision for a genetic override in emergencies, it would have resulted in much lower aggression levels and minimal social cohesion within the tribal group—a potentially terminal condition in those precarious early days. As it was, however, our growing sense of spirituality and an obsessive drive to attach mystical significance to our perceptions became the keys to the survival of our maladapted species. Here at last was a substitute for the fur, claws, and fighting teeth that evolution had failed to provide. Here was the magic weapon, the X-factor, that would launch *Homo sapiens* to evolutionary stardom.

The trappings of civilization that we see around us are not, therefore, the product of a rational animal but of a profoundly mystical one. If not, the inevitability of our present population explosion, along with all of its grim consequences, would have been painfully apparent to all educated people many centuries ago. Sharp-eyed observers of the human condition have, throughout history, noted that overpopulation invariably triggers the same grim sequence of social and environmental retribution. Finally two clergymen, Lütken and Malthus, became aware of the global scale of the threat and tried in vain to warn their contemporaries of it. Nationalistic paranoia, intellectual jealousy, and dogmatic religious beliefs—all good, genetically determined reactions—ensured that few understood, and none heeded, the warnings of those practical prophets.

Nothing has changed. Our genes still rule, as they should. Some still argue that humanity, with the aid of education, technology, and a new spirit of cooperation—resources unavailable to normal animals—will somehow manage to pull the survival rabbit out of the hat once more. Others, even less familiar with the true nature of biological evolution, believe that we are still progressing genetically. They argue that our growing intelligence, armed with better information, will eventually save us from the coming holocaust. Sadly, their faith is unfounded. Evolution is without intent and does not progress, and we are indeed normal animals in every sense.[28] Besides, both propositions fail to take into account a key fact of biological evolution: all species must fail eventually, especially the very successful ones, or the whole system would grind to a halt. In fact, the more genetically productive the species, the more important that it come equipped with some fatal flaw, some efficient self-destruct mechanism that will undo it in its hour of overwhelming triumph. This flaw usually resides in some distinctive physical or behavioral asset that has the capacity to turn into a liability when the environment changes the rules for survival. It now seems that even *Homo sapiens*, evolution's professional survivor, has failed to evade this ancient decree. Despite the astonishing behavioral flexibility that has steered this maladapted primate so adroitly through some 2.5 million hazardous years, the animal is still vulnerable in the way that all animals are vulnerable: through its adaptive specializations. By endowing the human brain with its language facility, evolution has ensured that human genes will continue to bypass the cerebral cortex at will, disguising fact with significance and turning imagination into perceived fact. This prodigious talent for spiritualizing its perceptions seems certain to keep this sapient primate safely sequestered from reality and well within reach of the biosphere's standard forms of population control.

There were three evolutionary prerequisites for our particular flaw: in view of our physical inadequacy it needed to be extraordinarily beneficial to begin with, and even when switched into its destruct mode it had to remain well disguised and thoroughly tamperproof. All of those evolutionary requirements have been fulfilled. Our time bomb is mysticism. Its delivery system is language. And its hiding place? The unfathomable coils of our DNA.

Chapter 9

Excalibur!

Why, then the world's mine oyster,
Which I with sword will open.
—*The Merry Wives of Windsor*, act I, scene ii

Meat-Eating Males

There is a tiny ball of ruffled feathers glaring at me through my study window, hopping from twig to twig and occasionally hurling itself at the glass in fury. My disgruntled visitor, a spotted pardalote (*Pardalotus punctatus*), is not of course glaring at me personally but at his own reflection in the glass. He perceives it to be a rival male and therefore a threat, both to his territorial dominance and to his family foraging for insects in the foliage behind him. Time and again his beak and claws crack against the pane until my concern for the safety of his brain, bashing against its bony case, causes me to draw the white curtains so that his reflection rival will disappear and his assaults will cease.

Despite a genealogy that incorporates almost 150 million years of spectacularly successful bird history and some 80 million years of dinosaur evolution before that, my poor punch-drunk friend is just as much prey to his emotion-loaded illusions as I am. His problem is not simply that he is a male predator with a family to defend. He also has a territory to defend. Territory is an emotion-loaded subject for most of us male predators, and with good reason: it is frequently a matter of life and death for us. In many species a male without a hunting territory is in for a short, unhappy life. Few females will mate with him and sooner or later he will die of starvation or be fatally injured by some outraged male property owner. Still dancing to the tune of those ancient genetic imperatives, like the pardalote, we are only too willing to switch reason off in order to hurl abuse, and if necessary, our bodies at any potential threat, however illusory.

Our ancestors would have discovered that the same principles applied to them when they began hunting on a professional basis on East Africa's hazardous plains 2 million years ago. A large, exclusively owned territory meant food, security, and genetic survival. They were faced with one of the oldest challenges in the evolutionary book: how to achieve these things in an increasingly threatening environment with nothing but a degraded and inappropriate defense system from the neck down. Bipedalism might have freed their hands for carrying food and weapons and given them the significant advantage of a higher viewpoint, but in that early ice-age world, where some of the herbivores were almost as daunting as the predators, a lone hunter clutching nothing more lethal than a lump of rock, a club, or a sharp stick would have had a short life expectancy indeed. To make up for the scarcity of fruit, nuts, and edible vegetation on those sparsely wooded plains, our ancestors would have had to invest more and more time hunting and scavenging for protein. This would have placed them in direct competition with full-time hunters and scavengers that were not only larger, faster, and stronger, but vastly better armed. With the odds so stacked against them their only option was to hunt or scavenge in tight-knit well-disciplined groups.

Judging by the way modern chimpanzees cooperate during their hunts for red colobus monkeys, a certain talent for collaborative hunting was already well embedded in the family genes. Nevertheless, a little amateur hunting of small, defenseless arboreal prey is a far cry from becoming a professional on East Africa's dangerous plains. In fact, those who did not take on the hazardous role of hunter-scavenger with inordinate care (and preferably some relish) would have been swiftly weeded out by their thoroughly professional four-legged competitors.

Who Shares Wins

Studies of modern hunter-gatherer societies have revealed another, more subtle inducement to hunt—and to hunt well. The hunter not only gains social and political leverage, but he also launches his genes more often and more widely. As anthropologists Kristen Hawkes and Jared Diamond discovered during their independent studies of modern hunter-gatherers, the successful hunter invariably shares his kills ostentatiously, even ritualistically, with all members of the tribe, rather than just his wife and immediate family. His enhanced reputation and status then leads invariably to more frequent extramarital sex, with the inevitable genetic consequences. Hunters with the best reputations are the most sought-after sexual partners and invariably leave the biggest imprint on the tribal gene pool.[1] It is a prospect the male gene cannot refuse. The business of hunting is thereby linked to the most powerful driving force of all and its lure would have been built into the male genome very early in our evolutionary history.

Nevertheless, it was not just the development of the sex-related hunting incentive that would have launched *Homo erectus* on its meteoric rise through the predatorial ranks, but rather the raft of very public social consequences that flowed from the change in diet. Hunting is usually a relatively unproductive use of time, and the bulk of a tribe's calories were (and still are) supplied by the gatherers of the tribe, the women. This division of labor, in which women foraged near the campsite while males hunted farther afield, not only guaranteed that the tribe ate regularly, thanks to the women, but that the whole group congregated each night to share the spoils of the hunt, thus enhancing the tribe's safety and significantly strengthening tribal bonds. Under these arrangements, more than half of the animal protein eaten by each tribe member, hunters included, was donated by other members of the tribe. Consequently, an outside threat to any one of them automatically represented a very personal threat to all, ensuring that each individual could count on the group rallying to his or her aid in time of threat. Loyalty to one's tribe remains a sound code of behavior for human beings and well rewards those who play the tribal game according to the book.[2]

Excalibur Aloft

The main reason humans became hunters was, of course, to obtain meat. Meat delivers more calories per kilogram than plant material, and this concentrated energy is more easily extracted than plant energy. The net result was that the time spent hunting and gathering would have dropped sharply whenever game became plentiful. Good hunters would have found themselves with time on their hands—time for making new and better weapons, devising improved hunting strategies, honing political skills, and especially making sexual use of the additional status hunting success bestowed. But above all, it would have given our ancestors the opportunity to exercise their rearranged vocal cords and enlarged Broca's area while pursuing their age-old pastimes of sex, sport, politics, and war. The development of rudimentary language would have bestowed on them an ability to infuse their surroundings, even their own existence, with supernatural significance—an ability that would ultimately cut humans adrift from the rest of the animal world, separating them even from other hominid groups. In this manner, with the aid of their enlarged brains, loosened tongues, and bloodstained hands, *Homo erectus* began to unsheath an evolutionary weapon of unparalleled power. Indeed, it would prove to be a magical sword, an Excalibur, that would transform this inept and insignificant primate into the most formidable animal the world has ever known. Moreover, this astonishing weapon seems to have fallen into our ancestors' hands at a crucial moment in hominid history. Where once there had been up to five branches on the hominid family tree, by 1 million years ago all were extinct save those communicative meat eaters, and they had dispersed with explosive speed halfway around the world.

Despite a quadrupling of our rational cortex over the past 2 million years, our genetic behavior remains on a hair trigger and our passions flare with primal ferocity. Consequently, human existence has always been characterized by two streams of endeavor, one rational and one mystical, or to be more precise, one cortical and the other genetic. The ultimate product of this curious union was civilization. Given its mixed parentage, perhaps it was inevitable that the offspring would be more than a little unstable. Civilization may bear the superficial stamp of both sides of the family, but it is clearly the genes rather than the cortical neurons that have generally dictated the ebb and flow of cultural fortunes throughout history. Territorial greed and ambition, racial prejudice, religious fundamentalism, and the corruptive processes of acquiring and wielding power— these are the things that ultimately determine the fate of nations, not wise counsel and clever gadgets.

Hallmarks of Humanity

The world is littered with evidence of this genetic dictatorship in the form of crumbling tombs, palaces, and places of worship—to the point that it would seem that the human animal pursues its mystic visions of good and evil, truth and beauty, divinity and immortality with all the dedication that other animals reserve for the mundane business of finding food and reproducing. Our habit of attaching mystical significance to almost everything that plays a role in our lives proclaims its genetic origins as loudly and as clearly as our use of complex language announces that we are

Stonehenge, southern England. Some of the smaller stones of Stonehenge were quarried in the Preseli Mountains in southwest Wales around 5,000 years ago. If they were transported mainly by water, which is most likely, it would have involved a journey of at least 380 kilometers (almost 240 miles). The larger stones, weighing up to 42 metric tons, were maneuvered overland on sledges and rollers from the Marlborough Downs some 30 kilometers (about 20 miles) away. It represents a massive genetic investment in the adaptive device known as culture.

members of the species *Homo sapiens*. Belief in astrology, reincarnation, witchcraft, creationism, voodoo, spiritualism, psychic phenomena, UFOs, and fairies, not to mention our traditional deistic faiths and the twin concepts of Good and Evil, are the real hallmarks of humanity. However flimsy and irrational some of these concepts might appear at first glance, we would do well to remember that most of them have withstood an onslaught of fact that should have utterly demolished them had they been based on intelligence rather than DNA. The secret of their persistence lies in what they give us in return for our belief.

All modern cultures still incorporate the social behavior patterns that once contributed to the evolutionary success of our hunter-gatherer ancestors. The forms of observance may vary regionally, but the overall pattern is universal, involving a rigid system of loyalty, honor, and discipline and incorporating some form of sanctified public marriage to reinforce the fragile emotional bonds between breeding pairs. Cultural beliefs also tend to foster a healthy suspicion, if not hatred, of strangers—especially those who might pose a threat either to the tribe's food resources or to the genetic assets embodied in its women and children. Since the survival of hunter-gatherers was wholly dependent on the ecological health of the tribal territory, their belief systems also incorporated a complex code of behavior that included respect for their environment and practical prescriptions for its conservation.

Australia's Aborigines epitomized these principles. Spirituality and mysticism governed every aspect of their lives, all day, every day, from birth to death, and it helped to make them the most accomplished hunter-gatherers the world has ever seen. For the past 10,000 arid years especially, their survival depended on adhering rigidly to well-proven, mystically sanctioned behavior in dealing with each other and with the fragile ecologies that surrounded them. In other words, a hundred thousand years of Darwinian selection had fostered the most genetically productive strategies for survival in Australia's fragile ecologies and then institutionalized these patterns in the inhabitants by fitting them with a cultural straitjacket of appropriate customs and beliefs. In this way our gods have always served as the high-profile executive arm of our shrewd but secretive board of genetic directors. With such experienced directors, divine middle-management, and devoted employees—how could we fail! But what of the looser grip on reality that implicit faith demands?

Faith: The Ultimate Weapon

Curiously, it is irrationality itself that serves as sword and shield in the hands of the true believer. The more unsupported and untestable the proposition, the greater the leap of faith required to grasp it. Consequently, the greater the satisfaction it engenders in those who safely make that daring leap, and the more energy and passion they expend in order to protect their investment. The 1940s film director Jacques Tourneur, who specialized in thrillers that hinged on supernatural events, shrewdly observed, "the less you see, the more you believe."

Although the secular advantages to be gained by evading the censorship of reason in this fashion may be few and suspect, faith conceals a genetic lure that in most circumstances is totally irresistible. The bait is an immediate return to our mystical womb. With one bold act of submission—and a quick retreat from reality—we can once more become members of a small embattled tribal group, mystically bonded to one another and inherently heroic: "We few, we happy few, we band of brothers"

(*Henry V*, act IV, scene iii). The urge to belong to a tribe remains incomparably seductive and thoroughly immune to rational accounting. Not only does it bestow a satisfying sense of identity and tribal exclusivity, but it also lights a bonfire of belief and self-righteousness that sends all those shadowy self-doubts scurrying for cover. Blind faith is indeed a movable feast as far as our genes are concerned. As some wit once observed, faith is the magic ingredient that enables us to make "that wondrous leap from grim reality into the totally bloody ridiculous."

So pressing is the need—in males especially—to achieve this euphoric state through brotherhood that warrior groups are driven to invent all kinds of social devices to reinforce the tribal bond. Novitiates are often required to adopt a bizarre mode of dress, or to have their bodies painted, tattooed, or otherwise engraved with symbolic designs, or even to undergo ritual mutilation to signify their servitude. Fear and pain, both mental and physical, have always been valuable tools in our genes' struggle to lock us into their secret agendas. And it matters little whether they are administered ceremonially during public rites of passage or secretly in school yards, alleyways, or factory toilets. There is nothing like a little honest terror to flood the cortical synapses with enough neurotransmitters to thoroughly numb the rational cortex. It prepares the novice for that exhilarating leap from the narrow confines of reality into the welcoming arms of irrational faith. The bizarre and often sadistic initiation rituals common to so many male-dominated organizations offer perfect examples of our 2-million-year-old "warrior" genes at work.

The Logic of Insanity

A paradox remains. Mystical beliefs do indeed provide us with an awesome weapon, but like the pardalote we are often driven to risk exhaustion, bodily injury, and even death in the pursuit of mere fantasies. The loss of reality that such behavior routinely entails would seem at first to outweigh most of its advantages. And in view of our species' peculiar reliance on reason to provide the tactical and technological underpinning for our mystically driven lives, such a weakness in our rational armor would appear to represent a particularly dangerous evolutionary flaw. So why *does* it work so well? With this question in mind, it is worth returning to the scene of our genetic origins to look from yet another angle at the forces that shaped us.

As soon as the first traces of complex human language began to develop, tribal elders would have found themselves able to wrestle more effectively with the inexplicable—why sun, moon, and stars moved as they did, what thunder and lightning meant, what caused the seasons to change, what preceded life and what followed death. According to some authorities, rudimentary human language may have appeared as early as 2 million years ago, but precisely when the human imagination took its first mystical flight, or how the innate spark of human mysticism was first communicated, we shall never know.[3] One thing, though, we can be sure

Lightning Spirit, Namarrgon, found in Kakadu, northern Australia. Arnhemland Aborigines believe that thunder and lightning are made by the spirit of lightning, Namarrgon, with the aid of stone axes attached to his elbows and knees. The sparks that fly from his axes produce the spectacular displays of monsoon lightning and gaudy orange grasshoppers (*Petasida ephippigira*).

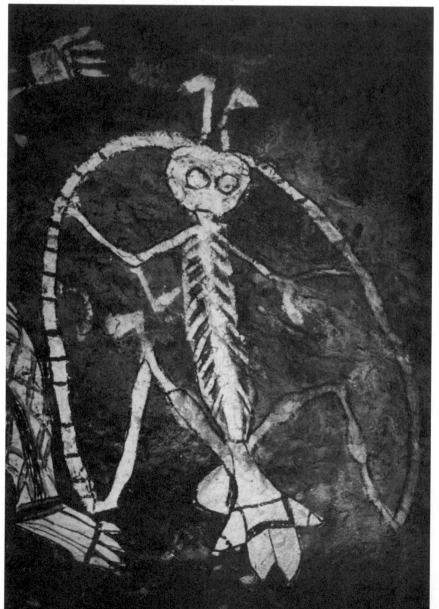

of: the spark that finally lit the bonfire of our modern mystical vanities came much later, and from a very real flame.

The domestication of fire was unquestionably the pivotal accomplishment in the history of human development. Its possession would have added immeasurably to tribal security, and in a sense it represented release at last from the thrall of nightly fear—the very real fear of waking in the jaws of a leopard, hyena, or saber-toothed cat. The campfire would have revolutionized family life in other ways too, extending the human "day" well into the night and thereby increasing opportunities for communication, food sharing, and storytelling. Along with its technological value as a tool for tenderizing and improving the taste of foods, straightening the shafts of spears and hardening their tips, and keeping stinging insects at bay with its smoke, the campfire by its very nature would have quickly become the focus of tribal life, just as it is with modern hunter-gatherers. All these factors would have contributed immeasurably to the strengthening of tribal bonds and the enrichment of the culture. It would not have taken the tribal mythmakers long to recognize its theatrical value. Old Aboriginal storytellers still know precisely how to wring the best effects out of the local seeds, grasses, woods, and foliage, punctuating their songs and dances with smoke, noise, and sparks.

The Eternal Flame

Cocooned though we might be within the comforting technology of our time, the flame remains our preferred symbol for human spirituality. Even now, fire still engenders strong emotions and lends wings to human imagination, whether it be the eternal flame that commemorates fallen heroes or the Olympic flame that athletes carry round the world every four years; whether the flames of purification used by the devout to expunge heretics and ward off evil spirits or by the ignorant to burn dangerous and threatening books. Even the flicker of a candle in a church, on a cake, or in a window confers peculiar power on its keeper and bestows disproportionate warmth on its devotees, a warmth that is inexplicable except that it comes directly from the campfires of our ancestors via the adaptive selectivity of Darwinian evolution. Those of our ancestors who were intrigued by fire and used it most effectively and most often would have tended to live better—and longer—than those who did not. And they would have left more descendants.

When exactly humans first learned to make fire and to use it as a tool is still uncertain. Remains of hearths and charred animal bones provide good circumstantial evidence for the controlled use of fire in Asia and

Europe as many as 400,000 years ago. But there is uncertain evidence for the occasional possession of fire that dates back almost 1 million years in Europe, and possibly as far back as 2.5 million years on the shores of Lake Turkana in East Africa. In other words, when humans first learned to gather embers from wildfires in order to feel a little warmer and safer at night, is anybody's guess.[4]

However, we should keep in mind that even *Homo habilis,* as an experienced scavenger, could not have failed to note that fires occasionally provided precooked meals and that birds of prey often used wildfires as a very effective hunting aid, preying on the wildlife that ran from the flames. With a 750-gram brain and an encephalization quotient (EQ) of more than 3.0 (see chapter 3), *H. habilis* would have possessed roughly the intellect of a modern six- to eight-year-old child, and it strains the imagination to believe that by 2 million years ago they had *not* learned to use fire, opportunistically at least. Boasting brains that weighed almost 1,000 grams by 1.5 million years ago, *Homo erectus* had an EQ that was even greater than *H. habilis* and could not have failed to do so.[5]

Apart from the feeling of power conferred on them by the possession of fire, our hominid ancestors would have had few other illusions about their lowly place within the larger scheme of things. Being wholly ignorant of the mechanisms that underpinned natural events, they could turn only to their imagination to fill the gaps in their knowledge and satisfy their hunger for explanations. It was inevitable that they gradually incorporated the entire fabric of their habitat—plants, animals, rivers, lakes, mountains, sun, moon, and stars—into their belief systems and wove them into their creation legends. These invariably entailed a list of behavioral responsibilities that became the contractual side of a very practical relationship with the environment.

Breaking the Faith

Such animistic faiths were to serve the human line very well indeed, until the demands of agricultural settlement finally ended that productive association and gradually swept the hunter-gatherer from the evolutionary stage. No elegies to the passing of the hunter-gatherer have been more eloquent than this:

> Every part of this earth is sacred to my people. Every shining pine-needle, every sandy shore, every mist in the dark woods, every clearing, and every humming insect is holy in the memory and experience of my people. The sap which courses through the trees carries the memories of the red man. . . .

The shining water that moves in the streams and rivers is not just water but the blood of our ancestors. If we sell our land you must remember that it is sacred, and that each ghostly reflection in the clear waters of the lakes tells of events and memories in the life of my people. The water's murmur is the voice of my father's father. . . .

All things are connected, like the blood which unites one family. All things are connected. . . . Whatever befalls the earth befalls the sons of the earth. If men spit upon the ground they spit upon themselves. This we know. The earth does not belong to man: man belongs to the earth. This we know. All things are connected. . . .

Where is the thicket? Gone. Where is the eagle? Gone. . . . The end of living and the beginning of survival.

Formerly attributed to Chief Seattle, leader of the Suquamish tribe of native Americans in the northwest United States, this poignant lament was supposed to have been part of a speech that marked the transfer of ancestral Indian lands to the federal government in 1855. That this statement has been liberally edited and rewritten several times and that these particular words probably bear little resemblance to those actually uttered by Chief Seattle is immaterial; they nevertheless convey with precision the feelings and sensitivities that once pertained to hunter-gatherers everywhere. Very similar sentiments were expressed to me personally by Aboriginal elder Bill Neidjie in 1978 and repeated publicly in the book *Kakadu Man*.[6]

Most Aboriginal songs show a similar bond with the natural environment at the most intimate and personal level. The following lines, translated by R. M. Berndt, are part of the Rose River song cycle from northeastern Arnhemland:

> [Mice] hopping along, shaking off the misty threads of the spider webs;
> Grasping the foliage and leaves of the new shoots and the swamp grass;
> [Mice] running to and fro, squealing and gossiping among the grasses;
> [Mice] running through spider webs, their bodies covered with mist and haze;
> [Mice] hopping along, leaving messages on their little tracks . . .
> Hopping along, grasping the foliage at the place of the Snake . . .
> Hopping through the bamboos, through the mist of spider threads . . .
> [Mice] leaving pawmark messages along their tracks . . .
> Among the swamp grasses, and among the new shoots . . .[7]

A Comprehensive Ignorance

A tide of civilization has all but extinguished the hunter-gatherer cultures that lived in such practical and sensitive intimacy with the natural environment. In contrast, Western society's long alienation from the natural world has bred an ignorance so comprehensive that it now represents a serious handicap in dealing with our environmental problems. Few have even the haziest grasp of the natural world that underpins their lives, or of the evolutionary processes that made them what they are; fewer still recognize their ecological place within the biosphere, or their ultimate dependence on it. A national survey of first-year biology students attending Australian universities in 1996 showed that up to 25% believed the biblical mythology of creation rather than the factually supported theory of evolution.[8] In other words, they preferred to believe that all human beings were the product of divine inspiration and were therefore separate from and unrelated to all other earthly life.

Despite—or perhaps because of—a lack of formal education, no tribal Aborigines are that ignorant. They might not countenance a genetic explanation as an alternative, but like all hunter-gatherers, they are far too intimately aware of the relatedness and interdependence of all species to tolerate such anthropocentric nonsense. A poll of medical students conducted in 1993 by Roger Short of Monash University, Melbourne, had even more serious implications for the community at large, showing that 27% of those future medical practitioners believed that species were immutable and had not evolved. Furthermore, 21% were unshakably committed to the biblical account of the creation of woman from Adam's rib. Similarly, a group of internationally linked polls conducted in 1995 indicated that even in this final decade of the twenty-first century, 55% of Americans still do not accept that they have evolved from other species. Nevertheless, one-third of them believe in aliens, ghosts, Big Foot, and the lost city of Atlantis.[9] Even in Australia, one of the world's more scientifically literate nations according to that poll, most people still prefer to believe that human existence is continually subject to supernatural intervention. In India, Russia, and the multitude of smaller nations dominated by dogmatic religions like Islam and Catholicism, such faith in supernatural powers is virtually universal.

The separation of our belief systems from their very practical origins began with the birth of agriculture between 10,000 and 12,000 years ago. It would have been a slow and painful process. Even though life in an agricultural community must have seemed much more secure and stable than the nomadic life of the hunter-gatherer, the powerful spirits that controlled the sun, moon, stars, wind, and rain still had to be reckoned with and reg-

Stations of the cross. A belief that supernatural powers intervene in earthly events is a definitive characteristic of human culture. The primary purpose of this carrot-and-stick device seems to be to reinforce genetic imperatives by generating powerful group bonds, encouraging altruistic behavior within the group, and identifying aliens and potential enemies.

ularly placated; the predatory forces of evil had to be warded off and the spirit of fertility coaxed into fecundity. So farmers and villagers took the only course open to them: they domesticated their hunter-gatherer deities, reinventing them in more manageable forms that could be carved, painted, or engraved on the walls of their buildings or carried about as bodily ornaments. As their beliefs became more and more icon-dependent and detached from the natural phenomena that spawned them, their faith became more open to reinterpretation. Released at last from its bondage to natural phenomena, the human urge to spiritualize was finally free to go cosmically feral and invent its own, more malleable gods.

A belief that supernatural forces manipulate events in the physical world is still a thoroughly definitive human characteristic, overwhelmingly prevalent among individuals and universal among societies. The traditional conclusion drawn from the ubiquity of such belief is that it proves the truth of the basic concept. A more obvious and less circular inference is that, like our predisposition to talk, our tendency to believe in supernatural forces is genetically based. But this simple explanation is usually avoided.

Modern Mysticism

For the most part, of course, humans are enriched and strengthened by their so-called spirituality—a fact that eloquently affirms its evolutionary value and the efficiency of the Darwinian selection process that embedded this characteristic in the human genome. Mysticism of one kind or another underpins virtually all of our literature, art, music, drama, legend, and law and provides paradigms of excellence that have been responsible for producing almost everything that our species deems to be of lasting value. In bedrooms and boardrooms, in theaters and classrooms, on playing fields and sidewalks, all around the world men and women wrestle on a personal level with concepts of good and evil, beauty and ugliness, love, lust, and hate. The Vatican, Las Vegas, Hollywood, Madison Avenue, Wall Street, the White House, fashion houses, and red-light districts everywhere literally depend on it for their daily survival. Nor is our penchant for the mystical confined to matters aesthetic, religious, or moralistic. It parades through the pages of women's magazines under the headings of astrology, palm reading, and psychic advice, and even hides behind the columns of figures that summarize the day's share trading, the football pools, and the local lottery as we follow our hunches and trade on our luck. The massive machinery of the world's multibillion-dollar gambling industry is similarly lubricated by the lure of good fortune as it manipulates the spin of the roulette wheel, the fall of the cards, and the roll of the dice. In gambling too it seems, we must believe the unbelievable in order to get excited. And wherever human emotions become aroused, mysticism is free to call the shots.

When reality strikes, however, the failure of faith may exact a savage fee. Psychologists, psychiatrists, and the gurus of spiritual development and self-fulfillment all find rich pickings among the hordes of disillusioned idealists, while packs of amateur counselors jostle for the scraps that fall from the professionals' table. In the shadows lurk less savory scavengers—drug dealers who promise brief but spectacular solace in elysian worlds fashioned from pills, powder, or smoke.

Stotting Males

Young males are especially vulnerable to such bait, since there is also considerable social advantage to be gained by the use of so-called recreational drugs. Much like African gazelles that execute a series of spectacular leaps high into the air when threatened, young men advertise their prowess by their risk-taking behavior. The gazelle's strategy, called stotting, appears to have evolved as a lifesaving demonstration of the gazelle's strength and fitness. It says to the predator, "I am too fast for you; don't waste your time in futile pursuit." And despite its risky and energy-sapping nature, the ploy usually works. The more impressive the stotting, the less likely that the gazelle will be pursued, and the better its chances of giving rise to similarly endowed offspring.[10]

Young would-be warriors use drugs for much the same reason. The more illicit and dangerous the drug, the greater the stotting factor. By this strategy their genes are saying to potential rivals, "Do not bother to compete with me; my genetic fitness is so great I am neither bound by convention nor damaged by these dangerous drugs; I am a leader." Although physically and psychologically destructive, and potentially lethal in the long term, this stotting behavior also seduces the user's genes with temporary delusions of superiority and invincibility. Thus, to many genetically insecure males, drug stotting appears to offer a very practical stratagem for establishing a broader beachhead in the next generation. The fate of the stotter is of little consequence. As far as our genes are concerned, reproduction is the only game in town, and once the new generation is launched the parent is expendable. Genes are neither wise, nor psychic; nor are they selfish. They merely play the percentage game according to million-year-old rules.

A second level of the drug world, occupied by those who engage in the even riskier behavior of drug trafficking, excludes those who merely use drugs. It offers the ambitious warrior a tight-knit tribal culture with its own peculiar language, rituals, and well-defined enemies. And if the warrior plays his cards right, dealing may afford him considerable wealth, tribal status, and mating opportunities. In some cases, membership in this priesthood is the primary goal and those who buy become objects of derision, considered less fit and less worthy than those who sell.

Computer Mythology

Oddly enough, some of the attractions of the computer culture are not so very different from those of the drug world. Many computer games offer instant entry to blatantly mystical worlds of sex, violence, and death, while beyond that lies the myth-ridden world of cyberspace, glittering with the promise of esoteric knowledge and enhanced tribal status. As in the drug world, the specialized jargon known only to the initiated adds to the sense of exclusivity and belonging. And lurking in the electronic shadows, like the dealers of the drug world, are the hackers. These renegade warriors satisfy their hunting instincts by conducting clandestine raids into the home territory of powerful electronic neighbors to loot or demolish their most valued possession—information. Enhancing the hacker's self-image, success may be used to impress his friends and improve his status within the tribal elite.

Conspiracy Myths

By far the most seductive and dangerous byproduct of our obsession with significance and the supernatural, however, is our devotion to the notion of conspiracy. Most of the injustice meted out by individuals or by society is clearly a byproduct of misguided zeal, ignorance, lethargy, or error on the part of others rather than the fruit of carefully orchestrated malice. Yet when we feel that we have been unjustly treated, our first response is to assume that there is conspiracy afoot and we are its target. Our inherent vulnerability to the lure of the conspiracy theory has ensured that this high-octane fuel drives most human conflict. As populations grow, and as the machinery of business and government becomes increasingly complex, so more and more people find conspiracy theories plausible and attractive.

So basic and seductive is this urge that thousands of otherwise rational white Americans feel compelled to join well-armed private militia groups expressly to defend themselves from what they perceive to be a corrupt and predatory federal government, or to train for "the coming race war" with black Americans.[11] Many African-Americans are similarly suspicious of white society and of the federal government in particular. Add this paranoia factor to the drug problem and it is hardly surprising that there are now some 250 million handguns in the country, one for almost every man, woman, and child. This was the kind of paranoia, directed against the federal government in general and the FBI in particular, that enabled the "Oklahoma bomber," Timothy McVeigh, to avoid rational consideration of the massive collateral damage to human flesh that he was about to inflict when he drove his explosive-laden truck to Oklahoma's federal building on 19 April 1995. The final body count was 168, with hundreds more injured. It was this same mind-numbing tribalism that set the crowd of vengeance seekers cheering in the street outside the courthouse when his death sentence was pronounced.

This explosive mixture of paranoia and lethal weaponry has led inexorably to a murder rate that is by far the worst in the industrialized world. As primatologist Frans de Waal points out, the homicide rate in Washington, D.C., is more than 34 murders per 100,000 people. By contrast, Berlin's murder rate is only 1.4 per 100,000, Rome's is 1.2, and Tokyo's is a mere 0.5.[12] All are bustling cosmopolitan cities yet overseas tourists who eat Washington's food, breathe the air, and drink the water don't seem to catch the bug, so the disease is clearly cultural in origin and is generally rooted in the delusion that the victim had a choice of behavior and chose wrongly.

Although this degree of detachment from reality seems at odds with a rise to world dominance that was so ably midwifed by science and technology, it must be remembered that the social soil that nurtured these dismal statistics was fertile indeed. Lured by dreams of Freedom, Justice, Democracy, and Wealth, the United States was colonized by the broadest possible spectrum of social and political idealists, religious refugees, and professional malcontents. Boasting such spiritual foundations, it is hardly surprising that it became the most overtly mystical nation in the industrialized world.

Given the right leadership and sufficient external threat, the primary product of such spirituality may be extraordinary social cohesion. This relationship is so predictable it has even been expressed in the form of an equation: Amity = Enmity × Hazard. Remove the external threat (Hazard) from ethnically diverse nations such as the United States and the former Soviet Union, and the internal divisions swiftly reassert themselves. Motivated by a volatile mix of disillusionment and paranoia, the warriors of each social fragment then reject national leadership and huddle about their local witch doctors, grunting with primal satisfaction as they hear again the old familiar tale of unseen evil closing in around them. With their primal hunger for tribal membership satisfied and a convenient enemy clearly defined, factual information becomes superfluous, even threatening.

The Primal Myth

Almost every leader of note has, either consciously or unconsciously, fished these murky waters at some time or other. If you were among the black-shirts chanting *Sieg heil! Sieg heil!* to your beloved Führer at Nuremberg on that September night in 1938, or if you huddled around the family wireless during Britain's darkest hours, trying to catch every nuance of those sonorous Churchillian phrases, you too have firsthand experience of the awesome power of the tribal leader who through intimate knowledge of his audience is able to tap the deepest veins of human mysticism. Their reward is a united people armed with humanity's shining Excalibur. To unsheath this magic blade, such visionary leaders must first win over the populace with the primal fairy tale, which invariably contains two ingredients:

1. A Monster—preferably one who speaks an alien tongue, prays to heathen gods, wears peculiar clothing, or has different-colored skin.

2. A Miracle—earned only by sacrifice, but culminating in triumph for the home team and a nasty end for the Monster.

This tired old routine has worked its magic with astonishing regularity since the dawn of history, and no one with fully functional DNA seems wholly immune to the lure of it. Its genetic nature shines through the grisly statistics that follow every major conflict, especially those that incorporate genocidal slaughter. They show with dismal clarity that it is not psychopaths that commit the bulk of obscenities, but the earnest accountant, the cheery carpenter, and that nice grocer on the corner. Such people have been responsible for most of the 25 to 35 million murders committed during the more notable genocides of this century.[13]

To achieve this miraculous transformation of ordinary people into genocidal killers, the grocer, carpenter, and accountant must first believe in a spiritual existence for human beings. Only then is moral distinction, and consequently hatred, available to them. Separate the mind from the body and all humanity becomes available for reclassification according to any mystic rules our genes care to impose.

Securely enshrined in our DNA, within every cell of our bodies, the ancient imperatives of human tribalism will no doubt linger on undiminished until our species breathes its last, but where tribal revenge might once have meant the death, by club or spear, of a neighboring family of nomads, the modern explosion of people and technology has put into the hands of the fanatic the power to deal out vengeance on a far grander scale. The practice of mayhem may not be much more common per capita than it was 10,000 years ago, but the precision of the steel blade, the speed of the bullet, and the wholesale efficiency of the gas chamber have brought an economy of scale undreamed of in ages past. Thanks to Joseph Stalin, Adolph Hitler, Chairman Mao, Pol Pot, and a few other gifted visionaries the twentieth century was the bloodiest of all. And still the crimson tide continues to rise, as tribally fragmented nations like Afghanistan, Angola, Burundi, Ethiopia, Rwanda, Sierra Leone, Somalia, Sudan, and Yugoslavia carve themselves up with obscene efficiency.

Genocide: A Family Trait

There is a widely held belief that "man's inhumanity to man" is a peculiarity of our species and that our tendency to genocidal behavior is unique in the animal world. We now know this to be untrue. The evidence was uncovered relatively recently by primatologist Jane Goodall and her colleagues at the Gombe Stream Research Centre in Tanzania. Painstaking surveillance of two groups of chimpanzees between 1974 and 1977 revealed a clear case of systematic genocide carried out by one of the groups. In a series of raids spread over the four-year period, well-disciplined war parties from the main Kasakela group finally succeeded in wiping out all members of a splinter group of seven males and three adult females, along with all of their young. On five separate occasions, distressed researchers watched groups of tense Kasakela raiders, in single file, silently closing in on members of the "traitorous" Kahama group. Having surrounded their victims, they then set on them with hands, teeth, feet, and whatever weapons came within easy reach. Although they were not lethal at the time, the injuries the raiders inflicted invariably proved fatal within a day or two. In several cases attackers tore flesh from their victims with their teeth, just as though they were dealing with a red colobus monkey (their usual prey). It seems they had mentally reduced the traitor group to the status of "lower animals," just as humans do. Clearly the "traitor" tag generates a peculiar degree of malice in both species, perhaps due to the fact that traitors inevitably share the tribe's genes, making them most likely to compete not only for the tribe's territory and food resources but for the tribe's females as well.[14]

Field researchers reported in 1996 that the Kasakela group had begun a new series of raids, this time into an area occupied by its other neighbors, the Mitumba group. Already reduced by disease from twenty-nine to twenty individuals, this group appeared to be threatened with the same fate as befell the Kahama chimps. Similar attacks have since been reported among other chimp groups.[15]

Females, especially the young and nubile, may sometimes be spared (only to be forcibly abducted and incorporated into the raiding tribe), but never the victims' offspring. This ensures a win-win situation for the victors' genes—the competition erased and the reproductive base of the home team enlarged, all in one hit. The same rules applied in Belsen, Bosnia, and Burundi. Statistics from the British conquest of the island of Tasmania illustrate the point precisely.

A Tasmanian Tragedy

When Europeans first began investigating its shores in 1642, Tasmania was home to some 5,000 Aboriginal hunter-gatherers, descendants of a group that had been isolated from mainland Australia by rising sea levels more than 10,000 years earlier. In sharp contrast to a dignified and cordial meeting with the notable French zoologist François Péron in 1802, the arrival of British sealers, soldiers, convicts, and other settlers the following year led almost immediately to fierce competition for the land and its resources, and from there to open conflict.[16]

The Tasmanians fought back valiantly, but ravaged by European diseases against which they had no immunity and armed only with spears and clubs, they stood no chance against the well-armed invaders. Looked on as vermin, the Tasmanians were exterminated by whatever means was readily available. Most clashes were bloody and brief, and mutilation and torture were commonly inflicted on the survivors in order to "infuse an universal terror" into the remaining tribes.[17] Males who were not shot, hung, or decapitated were often castrated. Women survivors were sometimes tied to logs and burned with firebrands, and sometimes turned loose with the heads of their freshly murdered husbands slung about their necks so that they would spread terror among the remainder of their tribe.[18]

Younger, more nubile women were often spared, primarily to provide sexual sport for the soldiers, convicts, and other settlers. Although small children were usually clubbed to death, some older ones were kidnapped to provide cheap, expendable labor on farms and in the settlements. A handsome bounty of five pounds was offered for the head of each adult Tasmanian brought in and two pounds for each child, for they were not easy to find. Consequently "black catching," as it was called, became big business. In one of the more notable encounters, four well-armed sheep herders surprised a large tribal group, gleefully shot thirty of them, and then tidied up by throwing the bodies over a nearby cliff. The site is still known as Victory Hill.[19] Many settlers were only too happy to contribute to the carnage with no thought of personal reward, whereas others killed reluctantly on the grounds that such measures were "in fact the kindest in the end."[20] Within thirty years the main Aboriginal population had been reduced to just seventy-two adult males, three women, and no children; with the death in 1876 of Truganini, the last "fullblood" woman, the Tasmanian genocide was complete.

Similar conflict characterized the early colonial history of most of the mainland Australia also, but because of the size of the continent, the inhospitable nature of much of it, and the larger Aboriginal population, it was a desultory, drawn-out affair with the last officially organized massacre (by police) occurring as late as 1928. Local Aborigines later estimated the death toll of that conflict to be at least three or four times higher than the official figure of thirty-one. The total number of Aborigines killed during this undeclared war is generally acknowledged to be something in excess of 20,000.[21]

Although this dismal litany of murder and torture might sound alien and obscene to most readers, the level of aggression and degree of malice seems to be entirely typical for the species whenever ill-informed individuals find themselves in conflict situations.

Peacemaker Genes

Genocide is the exception in the primate world, however, and most species, including humans, generally adhere to complex codes of moral conduct designed to ritualize major conflicts and limit the social and genetic damage they inflict. Most are settled by threat displays followed by submission and flight by one side or the other. Many instances of conflict settlement by a third party, usually a high-ranking male or female, have been well documented. It is especially common among chimps, bonobos, baboons, and stump-tailed macaques, but it has also been recorded in mountain gorillas, golden monkeys, capuchin monkeys, patas monkeys, vervet monkeys, and even red-fronted lemurs.[22]

In his book *Good Natured*, primatologist Frans de Waal notes that when a male troop leader intervenes, the peacemaking is remarkably even-handed and clearly aimed at preserving peace and order in the troop rather than advancing the tribal status of the peacemaker or his relatives and friends. Neither does peacemaking lead to reciprocal favors, such as food sharing or political support. In other words, other primates appear to be every bit as moral as humans. However, as de Waal and others point out, primate society is no more peaceful than human society. Thanks to a well-developed sense of revenge, the 30% casualty rate that is common among most species of higher primate is roughly the same as the casualty rate for the vendetta ridden hunter-gatherers of South America and New Guinea. Violent assaults are common among chimpanzees, and rape is frequent among the otherwise gentle orangutans. Even in peaceable gorilla society, secondary males often resort to infanticide to demonstrate their physical power and genetic potential. In most cases the bereaved mother then consorts and mates with the killer of her baby, thereby making infanticide a productive strategy for frustrated male genes. In fact recent statistics suggest that human rates of violence are comparable to rates of violence that characterize other great apes—with one exception: our closest cousin the bonobo, who consistently makes love rather than war.[23]

Human violence is unique, however, in one respect: although it occurs less frequently than in chimp society, thanks to our technology, our casualty rate is vastly higher. Consequently, this primate penchant for violence now poses serious problems for human genes. Although personal and tribal integrity must be maintained at any cost, rampant aggression may threaten the security and stability of the entire gene pool. So the world now watches the spread of genocidal conflicts with growing horror—yet rarely steps in to stop it. That the United Nations so often fails as a peacemaker should surprise no one; the wonder is that the organization exists at all. We are indeed fortunate that somewhere in our DNA those primate peacemaker genes appear to have survived unscathed.

Genetic Accounting

The genocidal behavior of the Kasakela chimps not only reaffirms our common genetic heritage but also demonstrates that they, too, are mystically deluded creatures. Although their lack of language denies them the satisfaction of dressing up their genetic imperatives in the moral regalia that so satisfies their humans cousins, they too attach emotion-loaded mental labels to other individuals, perceiving some to be "good" and others "bad," and they deal with them accordingly—sometimes years later, as at Kasakela. What ultimately governs the level of violence in all animals—humans, chimps, lions or lice—is what works best for their genes. In any confrontation the level of violence that ensues is the product of an unconscious analysis in which each adversarial genome balances the potential genetic advantage against the risk. The killing of juveniles is far less risky and therefore vastly more common in the animal world than is the murder of adults. The advantages at both levels are plain.

Within every major group of animals there is at least one species in which infanticide occurs regularly, and among rodents, carnivores, and primates it is common. But the particular pattern and frequency of violence depends on a multitude of environmental and social factors. Infanticide works well for the genes of male gorillas (about one in seven gorilla infants are killed by males), but it does not work so well for chimps and orangutans, and not at all for bonobos—they can never be sure the child is not their own. Rape works very well for orangutans (between one-third and one-half of all orangutan copulations appear to be rapes carried out by subordinate males), but it does not work so well for gorillas and chimps and not at all for bonobos—for whom sex is always readily available. The main hobby for chimps seems to be engaging in brief bouts of domestic and political violence that are noisy but not lethal. The final cost-benefit calculations that generate these seemingly antisocial behaviors occurs, of course, at a genetic level, not a conscious one; otherwise they might not take place at all and any genetic advantage would be lost to the group as a whole.[24]

Language: The Jeweled Scabbard

The genetic engines that drive human behavior are the same as those that drive all the great apes, but we are deaf and blind to them by genetic decree. If we could turn off the mental sound track and watch the action objectively for even a moment, our chimpanzee heritage would shine through with perfect clarity. As far as our genes are concerned, the language we use to attain our ends is all sound and fury, yet it invests our behavior with a potency that might well be described as magical. According to the Arthur legend, Excalibur's elaborately decorated scabbard was the real source of the sword's magical powers, multiplying them tenfold. And like Excalibur's scabbard, language is the sheath that gives human mysticism its power.

We will, of course, continue to believe that we are uniquely sentient and that unlike all other animals we possess the ability to live wholly rational lives. This comforting fantasy was epitomized by the seventeenth-century philosopher-mathematician René Descartes when he proposed, "I think, therefore I am." However naive and inverted that proposition appears in the light of recent biochemical research, it represents a perfectly predictable delusion, even now.

Our enhanced facility for reasoned thought is unavoidably aware of itself and thoroughly anthropocentric, so it is naturally inclined to take far more than its fair share of the credit for our unprecedented evolutionary success. In order to maintain that pleasant myth, it must also discount and discredit the mental capacity of other animals. Accordingly, like most philosophers before and since, Descartes reasoned that "we should have no doubt at all that the irrational animals are automatons." And yet, as history shows only too plainly, it is not rational thought that most characterizes human existence but mystically motivated behavior. The rise and fall of cultures, for example, has always been primarily determined by the tides of human passion, not by the ebb and flow of reason. Logic merely devises the cultural mechanisms and technological hardware that underpins those successes and failures.

The Butterfly Effect

Rational thought proceeds much as we have always perceived, in linear, ratchetlike increments that are conveniently accompanied by a flanking body of related information upon which we draw at will. By contrast, our emotional (gene-driven) behavior is so complex and finely balanced that it may never be fully understood. In fact, it appears to be every bit as complex as the earth's weather patterns. And like the weather, it is more easily explained in terms of the chaos theory than by simple cause and effect. Certainly, particular behaviors cannot be attributed to particular genes, nor even to particular gene committees. In fact, meteorologist Edward N. Lorenz's chaos analogy seems to be eminently applicable to human behavior. Lorenz suggests that the wing beat of a butterfly in Iowa could, in principle, trigger a typhoon in Brazil. Sustained by a complex array of dynamic input factors, chaotic systems are characterized by hair-trigger responses to minor input fluctuations that then express themselves throughout the system in massive repercussions. Sow the wind, and if the system is truly chaotic, you may well reap the whirlwind. Thus the featherweight impact of apparently inconsequential information on our finely balanced genetic imperatives is sometimes capable of generating an emotional monsoon in an unrelated behavioral arena. Lorenz's butterfly effect ensures that our attempts to predict human behavior will remain roughly as accurate and practical as long-term weather forecasting.

Such behavioral inscrutability is inevitable given the awesome complexity of our biochemistry. Not only do large numbers of our 100,000 genes play minor roles in most behavior, but also it now seems likely that many of those genes are themselves multifunctional. Recent evidence suggests that the vast database of genetic information encrypted in the human genome is multiplied many times by the genes' ability to release proteins through the membrane of the secreting cell in the form of peptides. These protein subunits are then recombined outside the cell in alternative arrangements to perform different tasks in different parts of the body according to on-site requirements.

For example, a single precursor protein produced by a gene that codes for egg-laying behaviors in a giant marine snail (*Aplysia* sp.) contains sufficient cleavage sites to allow it to be cut into roughly 2,000 different combinations of peptides, each combination capable of generating a different pattern of behavior. Similar cleavage sites have been detected in many of the proteins produced by human genes, enabling them to be cut into subunits and selectively reconstituted in a similar fashion. One such precursor protein provides the components for a particular adrenal hormone that acts on the frontal lobe of our pituitary gland while at the rear of the same gland other subunits from the same protein provide the makings for an endorphin-like peptide—one of a group of neurotransmitters that are implicated in memory, learning, sexual activity, pain suppression, depression, and schizophrenia.[25]

Touch any one of our emotional trigger points and a rich soup of such gene-induced neurotransmitters will swiftly flood our cortical circuitry, attaching themselves to receptors located in or near the synaptic junctions that separate each neuron from its neighbors. Some synaptic terminals are excitatory, while others are believed to inhibit the activity of the target cell. Similarly, certain neurotransmitters will excite, while others inhibit the conduction of nerve impulses across those synaptic junctions.[26] Like traffic police these neurotransmitters govern the flow of electrical impulses throughout the brain, directing them into genetically acceptable patterns to the point that we become invested with uncontrollable emotions. Pitted against this powerful biochemical dictatorship, rational thought grinds to a halt, and hard, unpalatable facts transform themselves into inconsequential details.

Chemical Relationships

Perhaps the most spectacular example of the selective nature of human perceptions—and of our genes' ability to "switch channels" in the virtual reality device inside our heads—is displayed by the subtle changes in the neurochemistry of the brain as our genes manipulate us into, and out of, intimate relationships. Different suites of neurotransmitters appear to be associated with different states of emotional and sexual arousal, and they appear to lock us into the particular patterns of thought and behavior associated with them. According to anthropologist Helen Fisher and her colleagues at Rutgers University, there appear to be three distinctive physiological states involved in the male-female bonding process, which form three crucial planks in the behavioral bridge that enables our genes to make that quantum leap from this generation to the next in relative safety. The first plank can be loosely described as instinctive sexual arousal, or lust; the second, infatuation; and the third, attachment. They usually overlap a little, but each may exist independently, fully orchestrating the attendant sequence of thought patterns and behaviors by the release of its own particular set of neurotransmitters.[27]

The first plank, sexual arousal, is driven by hormones such as testosterone (from the testes and adrenal cortex) and estrogen (from the ovaries and adrenal cortex). On the far span of the bridge, the chemical bonds designed to keep a couple emotionally attached to each other long enough to properly launch a new generation seem to be founded much more on oxytocin and vasopressin, two hormones released by the nervous system, than on the sex hormones testosterone and estrogen.

Only the chemistry of infatuation, the center span in the genetic bridge, still remains elusive and ill-defined, according to Fisher, despite its being perhaps the most intriguing of all since it incorporates one of the most recognizable and memorable sensations human beings ever experience—falling in love. Once our genes have targeted a potential partner, the presence or absence of the target, or the faintest signs of their pleasure or displeasure, can cause involuntary mood swings that may lead to memories that last a lifetime or to clinical depression and suicide. Inevitably this phenomenon is most prevalent among adolescents and young adults, the age at which evolution requires the mating drive to be at its height and the rational processes to be most easily overridden. Significant also is the lack of gender differences in infatuation behavior, especially in the drive for sexual monopoly of the target individual and a hunger for their reciprocation.

Fisher's team has found strong evidence to suggest that this obsessive and unstable mental state is characterized by complex, chemically induced switching changes in the neuronal circuitry within the brain. A technique known as functional Magnetic Resonance Imaging (fMRI) enables us to record the minute regional variations in blood flow that commonly occur during the process of thinking as a series of electronic brain scans. By combining the videotape and the fMRI record of a subject answering personal questions, researchers are able to relate particular patterns of thought and emotion to particular patterns of brain activity. In this way, a number of love-smitten volunteers recently enabled researchers at the Albert Einstein College of Medicine in New York to pinpoint the structures and sites involved in the production of the feelings associated with infatuation. Although the precise chemistry is not yet known, Fisher's current research suggests that the neuronal switching changes recorded by the Albert Einstein College team are directed by a group of neurotransmitters called monoamines, which include dopamine, norepinephrine, and the ubiquitous seratonin. Deluded, obsessive, and unstable though this emotional stage may be, if we lacked the temporary chemical clamp of infatuation, then there is every possibility that lust would not bind us together long enough for the slow-setting glue of attachment to harden into a strong pair bond. And without a durable attachment between parents, the risks to their children automatically increase. Genes cannot afford to take such risks.

Malleable Memory

An ability to play fast and loose with factual information does not, therefore, necessarily betray an unusual level of ignorance or mental incompetence; it merely signals the presence of exceptionally strong genetic imperatives. And all of us demonstrate this same ability to some degree in certain circumstances. Our malleable memory is useful in that it allows the brain to fill in gaps in our memory surreptitiously, thereby keeping us happy in the belief that our version is the right one. Although our neuronal filing system is vast, it is also limited and so must confine itself to soaking up information that is readily accessible—perhaps because some of it is already familiar in some respect or because it touches directly or indirectly on the hair triggers of our genetic imperatives. On the other hand, the vast amount of surrounding information eludes our perceptions and disappears unrecorded into the past. When we try later to recall those events, memory regurgitates its censored, skeletal narrative and the brain fills in any disturbing gaps by raiding our vast filing system and inserting supportive material from other sources. The more genetically important the original event, the more pressing the need to censor, warp, or augment the memory; all the while we remain blissfully ignorant of the frantic back-room editing that ultimately provides us with our supposed total recall.

This creative editing process typically incorporates elements of memories left over from other events in our lives or from fantasies or dreams, or they may even consist of snippets that have been plagiarized from other peoples' narratives. Similarly, convenient memory contamination can be readily reproduced in laboratory conditions and is primarily responsible for a phenomenon known as the false memory syndrome. There is substantial evidence that many people have been persuaded by well-meaning psychoanalysts to "remember" entirely fictitious childhood experiences by such means. The psychoanalyst has only to keep "reminding" the client of the fictitious experiences, particularly during hypnosis or while the patient is under the influence of a psychotropic drug such as sodium amytal, for the imaginary events to become "real." Provided that the analyst-patient relationship is good, the context subtly coercive, and the patient's emotions sufficiently disturbed, even the most bizarre sexual and satanic fantasies can be insinuated into the patient's childhood memories. Nothing, it seems, is too far-fetched—incest, satanic sex, murder, even the eating of babies—for the mind to embrace if the setting for the autosuggestion is appropriately conducive. Such memories, like all our other mystical delusions, seem to be the product of a divided brain and an ill-informed and obsessively verbal left hemisphere. Under extreme duress a bewildered left brain will concoct whatever memory appears most likely to ease the immediate threat. Whether that memory is founded on fact or fiction is of little consequence to our genes; only their imperatives are worth considering.[28]

Pulling Our Strings

It is hardly surprising that our genes are able to manipulate us to this degree through our emotions. Emotions been around far longer than rational thought, and the ability to display them clearly would have been a vital asset during our slow and hazardous divergence from the rest of the great apes. Evidence of this watershed is not only enshrined in the wiring of our brains but also built into the muscles of our faces. All ani-

Juvenile orangutan (*Pongo pygmaeus abelii*). Just as human children enjoy making faces in a mirror, this three-year-old orangutan has spent a long time experimenting with facial expressions, sometimes checking the results with her fingers. Here she tries to watch her lips move.

mals use both their bodies and their faces to display basic emotions such as fear, aggression, and sexual arousal. In 1872 Charles Darwin noted in *The Expression of the Emotions in Man and Animals* that like all the great apes humans seem to be especially equipped for expressing emotion by the manipulation of facial muscles. No other animal group has as many dermal muscles (there are forty-two) devoted to the control of facial expression as we apes do. The expanded cortical surface of the human brain, however, offers us even greater flexibility and control than the other apes possess. More cortical area is assigned to the control of our facial muscles than is dedicated to the fine motor control of our hands (see chapter 8). Research by Paul Ekman and Wallace Friesen at the University of California–San Francisco suggests that this sophisticated combination of facial muscles and cortical input is able to deliver some 7,000 different facial expressions.[29] Much of that sophisticated equipment is hot-wired to our genes via the hypothalamus and the brain's other ancient core structures, thereby allowing us to express feelings facially without first thinking about them. This cortical bypass sometimes betrays us by revealing emotions we would prefer to conceal.

Human beings become emotionally aroused in much the same way as other animals, and for similar reasons. Likewise, the particular regions of the brain that generate our emotions, and the neurotransmitter substances that regulate them, are remarkably similar to those in other animals. This similarity extends far beyond the ape family and beyond even the mammals. Traces of "our" neural chemistry have been discovered in marine animals such as octopus and crayfish. Their brains are large and regionally specialized like ours and contain large amounts of the neurotransmitter serotonin. As scavengers, crayfish need a highly developed sense of smell to find their food, and they use neural pathways, molecular chemistry, and olfactory brain structures that are surprisingly similar to ours. The similarities are so close they provide persuasive evidence that these chemical and neural pathways evolved long before our vertebrate ancestors emerged from the sea. However, the particular brain structures that use these chemical and neural pathways probably evolved separately at different times and have converged in design because that is what worked best in each animal group.[30]

Passion's Pawns

Also left over from ancient evolutionary accomplishments are the mechanisms that produce and regulate our emotions, which are based in the group of ancient enigmatic structures such as the hypothalamus, the midbrain, the cingulate gyrus, and the amygdala that lie at the core of the human brain.[31] This is not to suggest that our emotional responses are in themselves primitive, inappropriate, or capricious; if they were, then the *Homo* genus would not have survived its precarious infancy. By contrast, it was for eminently sound evolutionary reasons that humans came to depend so heavily on emotions as the driving force of their lives, and on their face muscles for communicating those emotions.

When the very first human beings were struggling to find new ways of surviving on the drying plains of East Africa some 2.5 million years ago, the clear communication of their feelings by facial expression would have been invaluable in bonding the tribe and turning it into a formidable fighting unit. Possessing only limited vocal skills, the more tightly knit and easily aroused the tribal group was, the less vulnerable it would have been to assault by neighboring tribes or predators. Faced with a screaming band of rock-throwing fanatics, no right-minded leopard would press home an attack when it might profitably redirect its aggression to more predictable prey such as gazelles or baboons. Similarly, human intruders, especially males, posed a very serious threat to the genetic integrity of small tribal groups, and a touch of paranoia in those circumstances would have been an invaluable asset for the home team. Defenders who reacted together with disconcerting speed and disproportionate aggression would have stood a far better chance of keeping their hunting grounds and their tribe intact. In fact, the more irrational and savage the response, the more likely that the defenders' genes would continue to dominate the region. The home-team advantage still applies, but now it doesn't represent as valuable an asset as it once was. Two million years ago a heritable madness of this kind would have been a lifesaver indeed. As any football coach knows, if an average team can be welded into a disciplined band of blood-thirsty fanatics, even dragons are slayable. And once battle is joined, should things go badly for us, our genes still have an ace up their sleeve. They turn some of us into heroes.

There is no escaping the fact that heroism works, perhaps not for the hero but certainly for the tribe. Genes code for altruistic behavior because they are not confined to individuals but are scattered in allelic forms throughout the population. The long-term survival of any particular gene—or more precisely, the information it encodes—is intimately linked to the survival of the gene pool as a whole. Consequently, where the survival of the gene pool might be enhanced by the altruistic behavior of one or more individuals, then those individuals who are genetically suited to the task are made to perform whatever sacrificial acts might best achieve that end. In this sense, at least, self-sacrifice represents the most explicit expression of genetic selfishness.

Kamikaze: A Divine Wind

History's most spectacular example of this kind of genetic selfishness occurred near the end of World War II, when some 10,000 young Japanese pilots volunteered to hurl their planes, loaded with explosives, into the hulls of American warships in a last-ditch attempt to stave off Japan's inevitable defeat. Known as the kamikaze, or Divine Wind, this suicidal tactic inflicted savagely disproportionate damage on the U.S. Pacific fleet and seriously hampered the Allied effort to achieve a swift end to the Pacific war. The original Divine Wind was a providential typhoon that saved Japan from being overrun by a massive Mongol invasion led by emperor Kublai Khan in 1281. The outnumbered defenders had been on the point of defeat after fifty-three days of desperate hand-to-hand combat along a 160-kilometer (100-mile) front when the kamikaze struck. It so ravaged the Mongols and their support fleet during the next two days that less than half of the invaders ever returned to China.[32]

Millions of human beings have knowingly sacrificed their lives in the heat of conflict or when loved ones have been threatened, and the instinctual origins of such actions are entirely understandable to most of us. But for kamikaze pilots it was slightly different. Death was not just a probability attached to grossly unequal conflict; it was preplanned and certain. The commitment to suicide by the young men of the Kamikaze Corps was undertaken far from the scene of conflict and entailed meticulous preparation, considerable ritual, and perhaps an hour or so alone in the cockpit before the target came within sight. There is no mistaking the source of their spiritual orders: the genes of the samurai warriors of old had reached down through the centuries, awakened their modern descendants, and locked them in the iron grip of honor. These young men had no choice but to obey; survival of the gene pool was all that mattered. Here was genetic selfishness at its most naked and unadorned.

As a result, in the last ten months of World War II, more than a thousand healthy young men soberly bound their foreheads with the white silk that symbolized a samurai who had chosen to die in battle and began the ritual of sacrificial death. After prayers for their success, each took his final sip of sake, climbed into his stripped-down, explosive-laden plane, gunned the motor, and set out to rescue the empire from defeat and dishonor. That many of the 10,000 kamikaze volunteers did not make that final pilgrimage was due to Japan's lack of expendable aircraft by that stage in the war. And political surrender finally put an end to the social structure that fueled such genetic dedication.

With centuries of success—and isolation—behind it, the ancient samurai code of Bushido, including its ultimate dictum of death-before-dishonor, had become deeply embedded in Japanese culture. A byproduct of this fierce code was a deep distrust of all foreigners. During World War II, that distrust was translated into an intense propaganda campaign that portrayed Americans as the ultimate barbarians, worthy of the utmost fear and loathing. The fertility of the genetic soil on which this propaganda fell was such that during the U.S. invasion of Saipan a total of 15,000 Japanese civilians committed suicide in the three days following the collapse of military resistance. As U.S. marines advanced toward the northern tip of the island, they came upon more than 10,000 Japanese civilians jumping from the high sea cliffs. Fathers dashed their babies on rocks and hurled older children and wives from the precipice before jumping themselves. Even worse, some 75,000 Okinawan civilians ended their lives in a similar way rather than submit to capture.[33]

The Genetic Bottom Line

In the code of the genes no quarter is asked and none given. The genes are preserved at any cost, and extremists often preserve them best. That this antique genetic strategy still works, even in modern warfare, is borne out by the grim statistics of the Divine Wind. For the outlay of just 378 suicide planes and about 600 airmen, the Japanese killed 2,000 American seamen, sank sixteen ships, and damaged eighty-seven others. (At least another 220 died in escort aircraft.) At Okinawa, in the face of awesome naval firepower and overwhelming air superiority, 930 kamikaze pilots hurled themselves at the massive Allied invasion fleet with similar success, sinking an aircraft carrier, ten destroyers, and six lesser ships and significantly damaging twelve other aircraft carriers, ten battleships, five cruisers, and sixty-three destroyers. And more than 3,000 U.S. servicemen died in the process.[34]

But suicidal attack is nevertheless a strategy of last resort, even for our extremist genes, for the risks are also considerable. As early as September 1944 the initial defense plan for the Philippines was secretly looked on by the Japanese General Staff as one last glorious *Götterdämmerung* gesture before inevitable defeat. (Japan's larger ships did not even have sufficient oil to engage in a protracted battle, let alone return to their bases in Japan.) Unless a miracle intervened, the 300,000 servicemen involved were simply expected to sell their lives as dearly as possible—which they did. Similarly, the last of Japan's super-battleships, the 64,000-ton *Yamato*, was sent to its certain destruction in a kamikaze-style assault on the 1,500 ships of the Allied fleet lying off Okinawa during the second day of the April invasion in 1945. Even though the war was clearly lost many months earlier, Admiral Toyoda, who ordered the attack, signaled, "The fate of our Empire truly rests on this one action." He had instructed the *Yamato*'s commander to inflict as much damage on the enemy as possible, and when the vessel's destruction was finally imminent, he was to ground the super-battleship on Okinawan rocks so that whatever remained of her 2,000-man crew might swim ashore and join the doomed defenders in the island's southern mountains.[35]

The attack totally failed. Sighted early by U.S. carrier planes, the *Yamato* and her woefully inadequate retinue of one cruiser, eight destroyers, and a few smaller ships were set upon by a massive strike force of bombers and torpedo planes, and the pride of the Japanese Imperial Navy went down with all 2,000 crew, far out of sight of the opposing fleet. With her went the cruiser, four of the destroyers, and another 1,000 crewmen.[36]

Only a genetic explanation can properly account for self-destruction on such a scale. Here was genetic altruism at its most explicit. In such dire circumstances even the most rational of military commanders can be moved to make "that wondrous leap from grim reality to the totally bloody ridiculous," condemning thousands to unnecessary death in the process.

Here is author David Bergamini's moving summary of those grisly final months of Worlds War II:

> The cold, cold statistic is that, in all, 897,000 or one of every seventy-eight of the proud, emotional Japanese people died in those last nine months of the war. They died with animal ferocity and despair, buried in caves and trapped between infernos. They died full of hate and hungry craving: famished, thirsty, and febrile; broken, twisted, and maimed; soiled, suppurating, and loathsome to their own nostrils. Their American executioners—of whom 32,000 also died, and bravely—did only what the Japanese themselves seemed to demand before they would surrender.[37]

And that, of course, is the very stuff of war.

Our Well-Tested Genes

If such behavior still surprises, we should remind ourselves that human morality represents the disguised expression of genetic imperatives that have survived more than 2 million years of field-testing. Even now, strict codes of honor and social conformity are generally synonymous with genetic survival and evolutionary success, whereas nonconformity and uncontrolled self-preservation tend to be genetically unproductive. As the decay of human tribalism, under the onslaughts of population growth and galloping technology, pushes us further adrift from the disciplined lifestyle and rigid tribal structure favored by our hunter-gatherer genes, the less efficient our genes will be in preserving the gene pool as a whole.

By and large, however, altruistic moral behavior still works in modern society to some degree, and our genes will continue to force us to endure hardships and hazards—and occasionally death—in preference to dishonor. Our obedience to such time-honored genetic strategies hinges on the strength of the emotions they are able to arouse in us. Consequently,

Anti–Vietnam War rally, Perth, Western Australia, 1970. These are public expressions of a multitude of genetic bargains previously hammered out within each participant. The resulting clash of moralities and high levels of emotion signal that the rational cortex has been overridden and that genes have assumed full control.

the primary characteristic of moral behavior is emotional conflict, and it is in the resolution of such internal conflict that humanity finds its recipes for valor, honor, loyalty, and the self-sacrificial behavior, both public and private, that we prize so highly. I believe that such behavior is the public expression of complex bargains hammered out in secret by opposing gene committees. In such emotionally trying circumstances, however, spirituality invariably comes to our aid, towing its string of colorful Trojan horses—Love, Honesty, Truth, Justice, Loyalty, and Family Values. They provide a perfect disguise for behavior that our genes might otherwise find difficult to justify in the open courts of reason. Safely concealed inside these Trojan horses, our genetic imperatives are able to slip safely through the communication gap in our divided brain and enter the behavioral arena intact and unchallenged. It enables "Right-to-Lifers" and Roman Catholics to champion the birth of each unwanted child under the Sanctity of Life banner while entire species become extinct at a minimum rate of three an hour under the present overload of human beings.[38] Meanwhile the emotional heat of the debate only too clearly proclaims its genetic origins.

Hungry as we are to assume the mantle of spirituality that supposedly defines our species, we tend to forget that precisely the same basic rules of existence and behavior apply to all vertebrates. The only difference is that they need no Trojan horses to breach the defenses of an analytical and censorious cortex, so their delusions are much less elaborate and pretentious. I'm sure my spotted pardalote needs no excuse for attacking his reflection, however irrational that action might seem on impact. The poor human zealot, on the other hand, whether Irish, Bosnian, or Lebanese, must feed his obsessive hatreds with a constant diet of righteous passion and gaudy imagery. It is a genetic sweatshop in there. No holidays for guerrilla genes.

Inevitably, most of the irrationally aggressive behavior exhibited by *Homo sapiens* is, for excellent evolutionary reasons, concentrated in the male. Two million years ago there was no alternative: the warrior had to be the male. Females were already fully occupied in nurturing their slow-maturing children. Besides, the male genome already possessed the basic genetics required to produce the necessary insanity. That is certainly the case with the common chimp, the orangutan, and the gorilla. Only the male bonobo managed to escape the dirty, demanding roles of warrior, status-seeker, and satyr and consequently eluded the debilitating derangements that inevitably go with them.

Gender Conflicts

Among the more common causes of conflict among all apes are those arising from the differing genetic imperatives of the male and female, especially in the arena of sexuality. Although the strategic aim of all genes is to inject themselves into descendant generations, the tactics forced on them by their different gender roles frequently bring males and females into conflict.

In primates, as in other mammal society, the male is best served by a wide distribution of his sperm. Promiscuity and minimal parenting offers the male the best possible chance of producing viable offspring and thereby achieving a secure toehold on the future. Since only the female can conceive and suckle the infant, she inevitably falls into the role of primary caregiver and therefore has a different tactical agenda. She becomes sexually receptive only when she is neither pregnant nor lactating. Consequently, her genes are best served by bonding with a competent, high-ranking male who will not only provide her progeny with good male genes but also help to protect both mother and infant during those vulnerable years. On the other hand, should she become pregnant to a genetically inferior male, she irrevocably commits that entire nurture period—almost a quarter of her reproductive life—to the error. By contrast, if a promiscuous male makes a genetically unsound match, it matters little to his genes, since some of them will almost certainly make it through to the next generation via other, more prudent matings. The genome of the female chimp, therefore, has always had far more at stake in a liaison than has the genome of the male, and it was this that led to the evolution of their different sexual strategies.

When humans first began to diverge from the other hominids almost 3 million years ago, this ancestral discord between the sexes was magnified by a gradual slowing of the human body clock. Three crucial changes occurred: the gestation period gradually extended, human babies were born in an increasingly helpless state, and they took longer and longer to mature. (It is quite probable that the adoption of the upright posture and consequent narrowing of the birth canal may have filtered out female genomes that produced more fully developed babies from a prolonged gestation period. Only less developed ones could fit through the reduced aperture.) The genes of the human male now faced a far more serious dilemma. If he failed to form long-term bonds with his several sexual partners, and failed to invest time and effort to protect his genetic interest in their joint offspring, the odds against any of his genes surviving increased significantly. But he could not be everywhere at once. If, on the other hand, he committed himself to a single long-term pair-bond, he deprived his genes of the obvious statistical advantages offered by primate promiscuity. Evolution was squeezing the male where it hurt most. Not only does prostitution, pornography, and the world's advertising industry now blatantly exploit this ancient male promiscuity imperative, but also the voyeurism it engenders is responsible for much of the profit in the fashion trade in general and photographic modeling in particular, as well as the massive market in sexually oriented movies, videos, newspapers, magazines, and books.

Well-Tailored Derangements

To be fair, the human female also has areas of genetically tailored insanity, but these are predictably different and much less bizarre in their behavioral byproducts. Being primarily concerned with achieving and maintaining a long-term pair-bond in order to optimize her fertility rate, the genetic imperatives of the female rarely clash with the rational processes of the cortex. Consequently the behavioral profile of such imperatives is so low that they seldom need to be disguised by ritual or justified by morality. But threaten her children or her relationships and you court responses any male terrorist would be proud of. "Hell hath no fury" indeed. Similarly, television's daily soap operas feed the female's compulsive interest in relationships and help to make up for the lack of village gossip in Western culture. In the sense that they vicariously satisfy her gender-specific, primal genetic drives, soap operas are the female equivalent of blue movies.

The poor male, on the other hand, is continually forced into situations where he must resort to extreme forms of mystical behavior in order to smuggle his genetic imperatives past his analytical cortex. It is this constant need for genetic concealment that produces the spectacularly irrational behavior of male-dominated cultures and leads so often to physical aggression between individuals and between families, between religious, social, or racial groups, and sometimes between entire nations. It also causes the gross disparity between male and female representation in the power structures of the world, especially those most firmly built on mystical foundations, such as political parties, cults, and religions. The intrusion of large numbers of female genes into these male display arenas inevitably represents a significant threat to the males' free pursuit of their very different, though equally legitimate, genetic imperatives, imperatives that hinge on matters tribal and territorial and that require absolute concealment from the shortsighted censorship of reason.

Female "Circumcision"

Inevitably then, it is women who bear the brunt of this genetically based male derangement, and in male-dominated cultures women are reduced to the status of inanimate possessions that may be used, abused, and disposed of at will. There are some 80 million women and girls, mainly in Africa and the Middle East, who have had, or who are destined to have, the primary organ of sexual pleasure crudely sliced from their bodies, merely to satisfy the male urge for genetic security and tribal status. This operation, clitoridectomy, is performed each year on some 2 million girls in Africa alone, and in Djibouti, Eritrea, Ethiopia, Sierra Leone, Somalia, and northern Sudan 9 out of 10 women are mutilated in this way. The operation, usually described quite erroneously as female circumcision, takes three forms. The mildest and least common form involves the removal of the prepuce, or hood, that shields the clitoris. This is the only one of the three operations that is analogous to male circumcision. The second—and most common—operation involves slicing away both the labia minora and the clitoris, an operation that is roughly equivalent to the removal of the entire penis in the male. The third operation, politely referred to as infibulation, or if you are Muslim as "pharaonic circumcision," involves cutting away all the external genitalia and stitching up the two sides of the vulva to the point that only a tiny hole remains for the passage of urine and menstrual blood. In Africa the stitching is usually done with an Acacia thorn using silk, catgut, or horsehair, and the opening is preserved by inserting a sliver of wood or a small reed. The wound is then dabbed with whatever local custom dictates is the best healing agent, such as ash, herbs, or animal dung, and the girl's legs are bound together until healing occurs—or septicemia sets in.[39]

The advantage to male genes of the common, second operation is very clear: once the woman can no longer receive sexual gratification from intercourse, then her husband's genes can be more certain that no competitor will fertilize her eggs, occupy her womb, and gain representation in the next generation to the exclusion of themselves. The advantage of the other two operations, however, can only be in the exercise of male power and the advancement of tribal status.

Only culturally reinforced genetic imperatives could have devised such painful, dangerous, and counterproductive (for the female) practices. Victims of genital mutilation may become sterile or even die from trauma, blood loss, or infection. Yet so efficient is the cultural feedback mechanism that it has even enlisted the women as supporters. Having been sexually mutilated themselves, mothers of potential victims will often use the argument that clitoridectomy not only eliminates an inherently impure female structure, thereby minimizing the daughter's desire and opportunity for promiscuity, but that it also enhances her social acceptability and her prospects for an advantageous marriage. It all serves to announce that the ancient female imperative to bond with a high-ranking male lives on in *Homo sapiens*, entirely unhampered by reason. Female genes "know" from prehistoric experience that since males are necessarily voyeuristic, their own survival in succeeding generations hinges to no small degree on the physical attributes expressed in each of their female progeny. This genetic knowledge produces the universal obsession with female body shape and fuels the world's fashion and beauty industries. No male gene could write a more satisfying script.

"Unnatural" Behavior

Since all of these reproductive strategies are gene-based and therefore only marginally susceptible to cultural modification via education, we are fortunate indeed that the degree of control sought by male genes is moderate by animal standards. Conversely, none of the "unnatural" behaviors often described as unique or characteristic of our species are, in fact, either. Rape, murder, infanticide, pedophilia, homosexuality, obsessive promiscuity, and even necrophilia occur in other species—and the perpetrator is not always male. In fact, one of the more notable examples of sexual aggression is displayed by the female wattled jacana (*Jacana jacana*). Since it is the male bird that incubates and nurtures the chicks, he remains tied to the nest and faithful to his sexual partner throughout the breeding season. It therefore falls to the female to adopt the role of territorial proprietor and sexual aggressor, and females consequently compete with each other for mating rights, with the winner accumulating a harem of nesting males. However, should an unattached female manage to drive away a breeding female, the victor destroys the loser's eggs and kills all her chicks (since they bear her predecessor's genes). She then courts the resident male, and because he is both childless and single again, he usually mates with her. (Similar behavior has since been recorded in other bird species as well as other jacanas.)[40]

In the desperate competition to leave descendants, the genes of the female jacana dictate that she must routinely behave more ruthlessly than all but the most deranged of humans. It serves to illustrate most eloquently that sexual behavior is neither moral nor immoral, nor even gender-based. It is like all behavior, entirely gene-driven and situation-dependent. Because it lies so close to the engine of life, sexual behavior more than any other requires absolute protection from rational thought. In humans, that protection is afforded by emotion and reinforced by the genetic feedback loop of culture.

Fortunately, however, we humans are generally unable to see our behavior in the cold light of genetics—and neither should we. If the veil of our fantasies were to be rudely torn from our eyes, the dynamos of cultural life would be instantly disengaged and humanity's eternal flame would sputter out. So our genes have ensured that by and large this cannot occur. We therefore have no choice but to cling to our mystical concepts of good and evil and continue, like the pardalote, to fling ourselves at mere reflections in the name of God, Truth, Justice, and Honor, suffering in consequence the physical and mental bruising that such actions entail.

As the rising tide of population gradually submerges the remnants of our ancient tribal structures, and tears down their comforting, confining walls, perhaps we might be forgiven for occasionally feeling that our X-factor, our drive to attach mystical significance to almost everything, has become more trouble than it is worth. Our old Excalibur, so long the defender of our species, now wounds friend and foe with equal dexterity and makes populous and socially fractured nations virtually ungovern-able. This might at first seem to represent a sad finale for one of evolution's most majestic and inspired inventions. But perhaps this judgment too is typically anthropocentric and shortsighted. Might these not be signs that our rusty Excalibur is even now engaged in its final evolutionary task: to turn on its owner and help put a Gaian end to the human plague it helped to launch 10,000 years ago?

Chapter 10

Midnight Prognosis

Stand still you ever-moving spheres . . .
That time may cease,
and midnight
never come.
—Marlow, *Faustus*

The Mystical View

Human welfare, gauged by any of its traditional indicators, is generally improving. A greater percentage of the world's people are now better fed and housed and are more likely to live past middle age than ever before, despite population growth. This, economists point out, is primarily because human ingenuity and human technology have more than kept pace with the population explosion, and they give every indication of continuing to do just that.

Environmentally harmful practices, too, are under continuous assault all around the world as never before. In developed nations the air is cleaner than it used to be, the water less polluted and more drinkable, and food supplies are reliable, affordable, and more diverse. As politicians, economists, and other mystics hasten to point out, there is no irrefutable evidence that global warming is attributable to human activity, or even that it is altogether "a bad thing." Around 100 million years ago, for example, the world was much warmer than today, and this warmth coincided with a major flourish of genetic diversity. Such evidence leads many to believe that the long-predicted environmental crisis is in fact no more than the hobgoblin of fanatics and a propaganda weapon that has been cynically misappropriated by opportunistic lobby groups. They contend that environmental doomsayers draw their dismal conclusions from inadequate and suspect evidence selectively extracted from the confusing deluge of data now available. Others remind us that "the good old days" were "good" only for a fortunate few, and there is solid statistical evidence to support this view. Life is indeed less brutal, more leisurely, and less hazardous for a larger percentage of the global population than ever before, and this trend toward human well-being shows little sign—in the marketplace at least—of reversing in the near future.

If your view of the world is wholly anthropocentric and confined to the short term, then such arguments will appear thoroughly sound and absolutely convincing. But should you be passably acquainted with the mechanics and history of biological and geochemical evolution on this small and finite planet, such reasoning will appear to miss the point entirely. For example, large numbers of happy, healthy mice are essential during the growth stage of a mouse plague in order to achieve an evolutionarily "satisfactory" outcome to that plague—catastrophic collapse. If individuals did not thrive until very late in the plague, it could not continue to its natural climax and the population would not then crash. From an environmental point of view that would be disastrous indeed, because overly "successful" species, like mice and men, would dominate to the long-term detriment of biodiversity.

So, if human well-being is not a good guide to the security of our tenure of this planet, how will we recognize that our particular plague has almost completed its growth and is about to enter a collapse phase? Is evidence of environmental deterioration, however suspect and subjective, the only guide we have? And are we justified in suspecting that phenomena such as global warming and ozone depletion are now incapable of cure by a little moderation and moral rectitude on our part?

The Chaos Principle

To find answers to these questions let us explore in a little more detail the proposition known as the chaos theory. The theory, proposed in 1963 by meteorologist Edward Lorenz of the University of Massachusetts, suggests that there is order to be found in such diverse and apparently disordered phenomena as the world's weather, phyllo pastry, dripping faucets, and the beating of the human heart. According to the theory, in a chaotic system minute disturbances within neighboring cyclic patterns are amplified to the point that they diverge with exponential speed, despite similar conditions in the original cyclic pattern and the neighboring one. Since all earthly systems are ultimately finite, this accelerating divergence forces the neighboring cycles eventually to fold in on one another—like the isobars in a weather map or buttered layers of phyllo pastry. When this folding happens repeatedly, it produces a fractal object the structure of which, at any level of magnification, resembles the whole. Lorenz described both the fractal nature and the extreme sensitivity of such systems as the butterfly effect—in principle, the wing beat of a butterfly on one side of the world could trigger a tropical monsoon on the other, with the vortices of the monsoons echoing in macro form the micro vortices left by the passage of the butterfly.

The rest state of such a system is not order but "nonrandom disorder" in which the system is wholly dynamic, continually changing, and fluctuating in a highly irregular but essentially nonrandom fashion due to the vacillating input of a multitude of chaotic feedback loops embedded within it. According to Lorenz and others, chaotic systems are not only astonishingly sensitive and responsive but also surprisingly resilient and self-regulatory over the long term, and ultimately very robust. This is precisely why the human heart is both tough and reliable and yet sensitive to the merest flicker of emotion. Two Boston research teams who analyzed the fluctuations that characterize the rhythms of healthy and diseased hearts found, at both the macro and micro level, that health appeared to be synonymous with chaotic disorder, and disease with simpler, more orderly heartbeat patterns. According to Chi-Sang Poon, researcher at the Massachusetts Institute of Technology, "Our findings suggest that the healthy heartbeat is highly chaotic, and the rhythms of the unhealthy hearts were significantly less so."[1]

If the nonrandom, self-regulatory principles of chaos do indeed echo throughout the biosphere at every level of magnitude from the molecular to the planetary, then it is no longer possible to avoid associating these processes and principles with the concept of a self-regulatory Gaian planet.

To return to the original question then: do we have cause for serious environmental concern? If we assume that the biosphere is in fact the self-managing chaotic system it seems to be, then the answer is absolutely unequivocal. Indeed we do have cause for concern. As we have seen, the rest state of a truly chaotic system is not stability but orderly chaos, and the apparent stability we see is a fragile, momentary thing, which once disturbed tends to disappear with exponential speed, never to be recaptured. The hope that the environment might be coaxed back to stability and health is entirely unsupported. In fact, the biosphere is much more likely to respond with unpredictable and seemingly disproportionate violence to the disturbance already inflicted. As far as evolution is concerned, this is no bad thing of course. In fact, this inherent environmental instability is the only mechanism by which the earth is able to manage its biosphere on a long-term basis. As the principal destabilizing factor, the species *Homo sapiens* is a primary target for environmental retribution and given its very fragile food base, an easy one to hit. Most of the food species humans favor have been selectively bred for high productivity, not for hardiness in the face of environmental adversity, and most agricultural production is concentrated in low-lying coastal regions that will be the first to flood should sea levels continue to rise.

Supply and Demand

The definitive signal that an animal plague is about to peak and collapse is a growing shortage of food. Human food production finally lost the race against human fertility in the mid-1980s.[2] This was the first time in recorded history that production had failed to keep up with population growth, and the global food surplus has suffered several declines since then. As outlined in chapter 2, in the past decade the world's food supply has barely kept pace with demand. Wild fish stocks are in global decline, grain reserves have several times been reduced to less than two months' supply, and the agricultural acreage lost each year through urbanization, erosion, salinization, and other factors now more than matches the new acreage added to production. Having already overplayed our fertilizer-pesticide cards, our only hope of postponing the ultimate Malthusian crisis for two or three more decades now rests primarily on expanding the acreage under irrigation and on the ability of geneticists to bestow pest and disease resistance on crop species by manipulating their DNA. But as we have seen, both of these options are high-cost, high-risk strategies that are no more likely to provide long-term solutions than were the chemical fertilizers, pesticides, and high-yield species introduced in the 1960s and 1970s. But we have no other way ahead, and no means of retreat.

Meanwhile the two primary measures of biospheric stress, the rate of species extinction and the increasing carbonization of the atmosphere, continue to mount. With an extinction rate now running between 1,000 and 10,000 times faster than the original background rate of one species per million per year and a 33% rise in the level of atmospheric carbon in the past two hundred years, there is no mistaking the message—it is a unanimous thumbs down. Given the current shape of the human population graph (see figure 22), those indicators also spell out a much larger

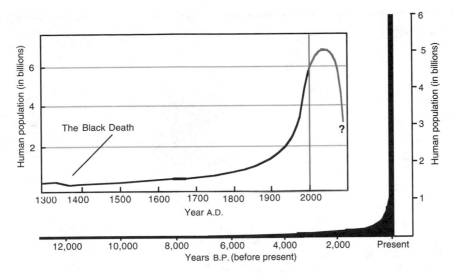

Figure 22. The human plague. The global population 12,000 years ago was about 4 million. Two thousand years ago it was still only a quarter of a billion. Two hundred years ago it had not yet reached 1 billion. The resulting graph displays the spike typical of all plague species.

and, from our point of view, more ominous message: the human plague cycle is right on track for a demographically normal climax and collapse. Not only have our genes managed to conceal from us that we are entirely typical mammals and therefore vulnerable to all of evolution's customary checks and balances, but also they have contrived to lock us so securely into the plague cycle that they seem almost to have been crafted for that purpose. Gaia is running like a Swiss watch.

Winding the Watch

If we were to assume for a moment the role of advocate for the biosphere, we should have to concede that what evolution most requires of us now is to ignore the warnings on the current environmental package and continue to behave as we always have. Should we falter in our rate of economic and technological progress we might jeopardize the nice swift end to the human plague that is now being signaled by the soaring extinction rate and the rapid carbonization of the atmosphere.

Unfortunately, our perspective of the road that led us to this population precipice is hopelessly distorted by our shortsightedness. Limited as we are by our own birth and death and the narrow span of existence that lies between, it is hard enough to imagine the passage of a thousand years, let alone a million. A billion is out of the question. It is therefore worth taking

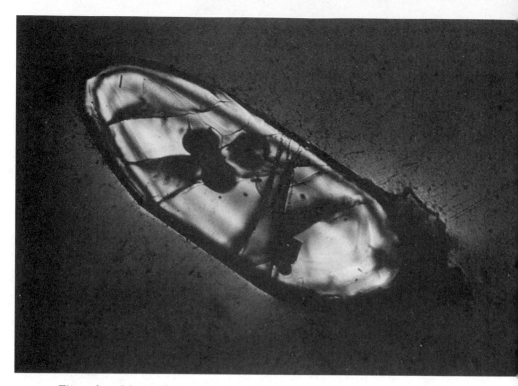

Zircon from Mount Narryer, Western Australia. This minute crystal of zircon is one of the oldest known fragments of the earth's crust. It began to crystallize almost 4.2 billion years ago, about 2 A.M. according to our twenty-four-hour timescale, and eventually became embedded in a piece of land that now forms part of the northwestern edge of Australia.

a moment or two to look at evolution and the emergence of our species against the larger perspective of geologic time, but in terms that we can more readily comprehend. The unit of time that most intimately governs our lives is the time the planet takes to complete one revolution about its polar axis—the twenty-four-hour day. If we overlay earth's existence on this scale and plot the milestones of its evolution against these twenty-four hours, we'll have a much better idea of the proportional time involved. For example, if we accept that the earth coalesces in the first few minutes of this day, then life's earliest trace fossils (the North Pole stromatolites mentioned in chapter 5), appearing with what has been described as indecent haste, show up just before 6 A.M. By contrast, the first multicelled organisms do not appear in the fossil record until about 9 P.M. on the scale, making it immediately obvious that the real miracle of life lies not in creation but in cooperation.

Dickinsonia costata fossil, 600 million years old, found in Ediacara, South Australia. The highly specialized cells represented here are one of the first multicellular organisms to leave a permanent impression on the seafloor. Since its precursors seem to have left no fossil trace, we can only guess at its evolutionary origins. Known as *Dickinsonia*, it may not even fall within the definition of an animal but rather may represent an antecedent group called *protoctists*. Yet against our twenty-four-hour scale, it does not make its entrance until about nine o'clock at night.

Our peculiar place within the evolutionary scheme of things is likewise illuminated by the application of the twenty-four-hour timescale to geologic time. According to our scheme, the first habitually upright apes did not begin to leave their footprints on East Africa's dusty plains until about fifty-five seconds before midnight, true humans emerged some twenty seconds later, and fully modern humans only made their appearance during the last four seconds of that final minute. It is an astonishingly brief existence for a species that in full plague mode now dominates the planet so completely that it threatens to destroy the environmental conditions that underpin its own survival. The dinosaurs, by contrast, tiptoed through time. They roamed the world for nearly an hour (some 180 million years) with an impact that was negligible by comparison.

So here we stand, at midnight, all 6 billion of us huddled together upon our small bank and shoal of time, with sea levels rising and an environmental storm gathering about us. The fate of our species hangs in the balance, and the matter will be decided for us within the next millisecond on our evolutionary stopwatch.

Evolution's Wild Card

Could we have changed anything in the past? Let's review those last fifty-five seconds and turn the clock back to the point when our apish primogenitor, a smallish hominid of chimpanzee stock, is trying to adapt itself to life on the drying plains of East Africa. A slowing of its maturation rate, and some minor modification to its brain—accompanied by rigorous Darwinian selection—is about to turn that hapless ape into an evolutionary wild card, an animal with little face value but almost limitless potential. In physical terms, however, this transformation will be bought at exorbitant cost. During those protracted evolutionary negotiations, our ancestors will be forced to surrender most of their original assets—thick body hair, long fighting teeth, and a great deal of strength and agility. These are very serious losses, only marginally offset by the retention of some useful odds and ends such as good stereoscopic vision, a pair of very mobile shoulder joints, and extraordinarily flexible hands. With an ice age looming, it is as if evolution is saying, "Come in chimp number three. Your time is up."

Even this primate's new features appear to be little more than cosmetic. A more versatile vocal sac, a fully erect stance, a prolonged childhood, and a semicontinuous sex life don't initially seem to lend themselves to ice-age survival. But concealed inside the bulging, misshapen skull of this fairly dismal package, there lies a solitary, glistening weapon the likes of which has never before been seen. In its general design the human brain differs little from that of the australopithecines. It is much larger, though, and the surface area is four times that of the australopithecine brain. This cortical enlargement, and the massively expanded neuronal circuitry that accompanies it, allows our predecessor to file, retrieve, and compare an astonishing volume of data. Perhaps most important, this enlargement has laid the neuronal foundations for the development of complex language, technology, and abstract thought.

In outward appearance, though, what a bizarre creature this is. Of great-ape stock, yet slow, puny, and nearly hairless, rather like an overgrown juvenile. Here is a prime candidate for the evolutionary scrap heap. And yet here also is the most dangerous animal of all—an armed fanatic. A "nonspecific" species, with no external weapons or defenses, yet a creature that seems to thrive on hardship despite its physical inadequacy. Here at last is the consummate survivor, the very paragon of animals.

When evolution launched its human prototype some 2.5 million years ago, the hominid family tree possessed three or four branches. By 1 million years ago (about twenty seconds to midnight on our twenty-four-hour scale) only a meat-eating species, our ancestors, remained. These hominids had not merely survived, however. Living and hunting in cooperative packs of twenty to forty individuals, they had already dispersed halfway around the world, swiftly establishing themselves as the top predator wherever they went. Their achievement is all the more remarkable when we remember that this venturesome hominid, *Homo erectus*, migrated from the drying plains of Africa into cool northern forests around 2 million years ago—just as the Ice Age unleashed its first major assault on the temperate zones of the world.[3] Armed with nothing more than sticks and stones, a little cunning, and a crucial streak of mystical madness, this extraordinary species gradually learned that by hunting in well-disciplined cooperative groups they could, with the aid of careful planning and a touch of fanaticism, bring down such formidable prey as the woolly mammoth, the saber-toothed "tiger," and the woolly rhinoceros. No wonder our genes still display the same magic blend of mysticism, cooperation, and technological competence that carried them safely through those hazardous early years. The question then arises: If those genes saved our hard-pressed *Homo erectus* ancestors, will they not rescue us in the same fashion?

I would argue that they will, but only in the sense that they will probably save us from extinction. Unlike the first environmental assault on our species, the present one is primarily anthropogenic and represents a biospheric backlash against our overwhelming success as a species. In anthropocentric terms, this one is directly aimed at us, and the very best technological defenses we can muster are likely to prove much less effective than the sticks and stones of our ancestors, for this time it is human technology itself that lies at the core of the problem. We are pitted against the full weight of the world's biospheric machinery, and to make matters worse, the carefully tailored mythologies that used to sustain us now handicap us with misconceptions that seem sure to be our downfall. To our peril we cling to the following beliefs:

1. Humans have spiritual autonomy and are therefore accountable for their actions.
2. The environment is inherently stable and will rebound if we give it half a chance.
3. Most environmental damage is the product of human ignorance, greed, and technological malpractice and is therefore repairable.
4. Human ingenuity and technology can solve most environmental problems and repair most environmental damage if given enough time, money, education, and political will.

5. We will survive, just as we always have, by drawing on our unique ingenuity, resourcefulness, and indomitable spirit.

By contrast, the facts are as follows:

1. We are genetically driven just like any other animal. We have no mind other than the body, and we lack behavioral choice.
2. The environment is a chaotic system and is therefore inherently unstable and always has been. If it were not so, evolution could not have occurred. Rebound is not characteristic of the system.
3. Most environmental damage is the inevitable by-product of overpopulation and is a necessary part of the plague cycle.
4. The environmental problems we now face do not have a technological solution. All human activity—"good" and "bad"—adds to our environmental debt. The more technological the attempted solution, the greater our environmental debt (Impact = Population $\times$ Activity $\times$ Technology).
5. The plague cycle is a vital component of the evolutionary process and an essential evolutionary escape clause in the case of a fertile, high-impact species like *Homo sapiens*.

Handicapped as we are by such monumental misconceptions, we are likely to discover that the endgame is rigged beyond redemption.

The Global Mousetrap

To be effective, a good mousetrap must incorporate a delay mechanism. The mouse must be given enough time to insert its body fully into the trap before the mechanism is triggered and the trap closes. There are two separate delay mechanisms operating on behalf of the global mousetrap. The primary source of delay is simply the size and complexity of most of the earth's biological and geochemical cycles. Having raised the volume of atmospheric carbon by a third, we should not expect the carbon cycle to adjust itself for at least another century or so—even if our greenhouse gas emissions cease immediately. Similarly, the oceans, covering some 71% of the planet's surface, contribute a massive thermal inertia that will absorb and conceal for decades most of the heat energy that is currently being retained by the carbon-loaded atmosphere.

Species decline is also slow to begin, difficult to measure, and impossible to stop. The removal of one or two key species in a fragile environment may lead to the extinction of a large number of dependent species, and

Methane bubbles, central Australia. Methane-producing bacteria feeding on decaying vegetable matter in the silt of a shallow pool in central Australia produced this glittering sheet of methane bubbles.

those losses in turn trigger other extinctions, producing a cascade of losses that may continue for centuries. Neither are the losses confined to the biological kingdom in which they began but may cross from plants to animals and back again, perhaps many times.

The second delay mechanism lies buried in our own brains. It is genetically based and resides first in our tribal nature and second in the seductive power of language. In the unkindest cut of all, Broca's area plays Brutus to our Caesar, distracting us with fantasies while the fatal thrust of evolutionary retribution finds its mark. One need only read the daily papers and watch the television newscasts to see the form this lethal diversion takes. What most readily claims our attention and occupies our minds is the behavior of other human beings, not the loss of species, the carbonization of the atmosphere, or the galloping degradation of the world's most productive land. We prefer that our threats be mystical and our monsters two-faced.

Meanwhile the trap itself is on a hair trigger. The mechanism is microbial and very well concealed. It is constructed from the shadowy second arm of the bacterial tree of life, the archaebacteria. These organisms commonly discharge methane, the most potent of the main greenhouse gases, and they tend to flourish in oxygen-free environments such as tropical swamps, peat bogs, and the permafrost of northern tundra regions. Should global temperatures continue to rise, the massive reserves of tundra methane will be unleashed and monsoon rains and rising sea levels will turn the world's tropical lowlands into methane-loaded swamps. The environmental repercussions that would follow would solve the population problem in just a decade or two. Frozen lakes in Siberia are even now releasing methane that has been locked in their icy muds for the past 27,000 years.[4]

We will remain oblivious to these threats, however, and like the other great apes continue to pursue the same old primate pastimes that have always preoccupied us: sex, crime, war, sport, and politics. Besides, even the gaudiest of fantasies may now be realized thanks to the wonders of digitized imaging, and all documentary images are consequently suspect and devalued. Mystics of all kinds are now free to run amok on a global scale, and the coming environmental backlash will provide all the proof they need.

Scaling the Peak

The obsessive urge to mysticise many of our perceptions is a mechanism that is eminently suited to its final function. All it must do is ensure that we continue with our political, economic, and spiritual enterprises, in which case we will pass the population peak and slip into uncontrolled collapse before we even know it. As the naive mythologies of economic rationalism tighten their grip on Western political thought and a tidal wave of industrialization sweeps through the populous nations of eastern Asia, the prospect of a global consumer culture based on technology and rampant consumption looms ever nearer, thus virtually guaranteeing that the human plague will continue uninterrupted until it reaches its natural climax in the middle of the twenty-first century.

According to the demographics of other plague-prone mammals, the rate of population collapse customarily matches the explosive growth that produced it. The biosphere is an impartial arbiter, and physically maladapted as we are, there is certainly no reason for us to expect special dispensation. If our plague does indeed match the traditional plague graph, our decline will mirror our explosive growth, and the population will halve itself twice in less than a century. Under such circumstances, the body count in the overpopulated countries of the third world will make the genocidal efforts of Hitler, Stalin, and Chairman Mao seem like the work of halfhearted amateurs.

Whether the mechanism that orchestrates our collapse will be a human version of the general adaptive syndrome (GAS) or whether it will be a fatal dose of environmental retribution is anybody's guess. The rodent version of GAS includes such physiological responses as glandular malfunction, inhibited sexual maturation, diminished ovulation and implantation, inadequate lactation, increased susceptibility to disease, and a sharp rise in infant mortality. The social aspects of rodent GAS include increased aggression, infanticide and cannibalism, curtailed reproduction, abandonment of unweaned infants, and a rising incidence of abnormal and unproductive sexual behavior, especially homosexuality and pedophilia. The reader can decide whether there is an analogy to be drawn. Levels of infant mortality are not rising and cannibalism has not appeared, but some of our current health problems, especially those linked to the subtle chemistries of the body's immune system, the autonomic nervous system, and the reproductive system, seem to be increasing, especially in developed nations. Many of these problems are commonly attributed, directly or indirectly, to chemical pollutants and are therefore considered to be symptoms of environmental degradation rather than innate physiological phenomena. Studies of other mammals, however, suggest that most plagues collapse in response to a combination of internal and external factors.

Our astonishing ascent from evolutionary obscurity to world domination has for the most part been slow and painful. From the impassioned howl of the primeval hunter to Beethoven's Ninth Symphony, from the persuasive grunts of a stone-age paramour to the spare elegance of Japanese haiku, from the painted caves of Ardèche and the rock shelters of Kakadu to the ceramic sails of the Sydney Opera House—these are milestones in an evolutionary journey that has no counterpart in the animal world and for which we have no adequate measure. Viewed from here, amid the glitzy clutter of Western technoculture, the traumatic birth of our species seems as remote as the stars; yet in an evolutionary sense it occurred only a moment ago and haunts us still. The shadow of a human hand stenciled on a cave wall was every bit as meaningful to our ancestors as the glowing computer screen is to us. And our most sophisticated laser-guided weapon is essentially no more than a space-age spear. We have merely traded the economy and elegance of the original for a bit more punch at the business end.

Meanwhile, the genes of our nomadic ancestors live on inside us, inviolate within the massive reservoir of human DNA. The evidence is everywhere, especially on weekends. Our tribal skirmishes might now be professionally conducted on carefully groomed playing fields, but the roar of the fans is bloody and primal, and any Australian park ranger will tell you how hard it is to prevent the urban warriors from drawing on the rocks, slaughtering the animals, and leaving the landscape in smoking ruins every summer weekend.

Urban Warriors

The very powerful gun lobbies of the United States and Australia are another eloquent expression of our ancient warrior genes. We may no longer live in the shadow of the saber-toothed cats, yet many of us still feel an indefinable attraction to the weapons that would best protect us if we did. Even Australia, a nation never overtly seduced by the gun culture that now bedevils the United States, waited until fifty-four people were shot, thirty-four of them fatally, by a solitary dysfunctional youth with a semi-automatic rifle before passing legislation prohibiting the ownership of such weapons. Even after the tragic shootings that occurred in Port Arthur, Tasmania, the law's passage was not easy, and despite the difficulties of logic faced by the gun lobby, its members succeeded in generating the largest public demonstrations (overwhelmingly male) since the anti–Vietnam War rallies of the early 1970s.

The highly emotional nature of the gun lobbyists' response could not have signaled its genetic origins more clearly. In hunter-gatherer societies of old, a warrior's favorite weapon was as personal as his penis and cherished accordingly. It is just the same for the urban warrior. Faced with a dearth of argument capable of justifying their genetic outrage, Australian gun enthusiasts, like their U.S. counterparts, were inevitably reduced to the last resort of all beleaguered genes: the conspiracy theory. This legislation, they said, was a poorly disguised attempt by power-hungry politicians to disarm the populace, especially those free and independent gun-toting spirits who would be first to resist a blatant assault on democracy and individual freedom.

Fortunately, however, the weapons lobby is generally outvoted in the finely balanced parliament of genetic imperatives that regulates our daily lives. And so it was in Australia's parliaments, although not without one brief but spectacular display of warrior petulance in Tasmania. There the male-dominated upper house introduced some unique amendments of their own to the state's new gun legislation, amendments that lowered the legal age for firearm use to twelve years and decreed that it was acceptable for licensed firearm owners to use their guns while under the influence of drugs or alcohol. Far from signaling an epidemic of senile dementia among members of the legislature, these bizarre amendments merely affirmed that their ancient warrior genes were still eminently capable of snatching the reins from the rational cortex whenever warrior pride was at stake.

The bloody slaughter in Port Arthur and the long chain of social repercussions that followed poignantly underscore the inability of a mere 10,000 years of agro-urban culture to override our 2-million-year-old human genes. The human brain remains a piece of stone-age machinery, however you look at it, and no amount of culture can make it otherwise. Nor are any evolutionary improvements likely to tip the balance. Genetically speaking we are a finished product, not a prototype. What you see is what you get—there will be no bright utopian future, just more of the same.

But this unhappy story has a less dismal footnote. As journalist Steve Meacham pointed out, although the Port Arthur killer was male, as is usually the case, five of his male victims died trying to shield their wives, and at least one mother did the same for her children. "In leaping into the path of a gunman's bullet they must have acted spontaneously, obeying some primal selfless urge to risk themselves so that those they loved the most might live," wrote Meacham. "In the split seconds their fate was cast, their reactions were based on reflex rather than reflection."[5] There was indeed no room for rational thought on that tragic afternoon, only good, honest animal behavior. The heroism of those six could not have been more explicitly genetic.

Evolution's Street Kids

We have an identity problem. By genetic definition we are hunter-gatherer chimpanzees, yet we no longer live in a social or physical environment that suits this time-honored role. By hunter-gatherer standards we have become evolution's street kids, rootless, aimless, and living by our wits. We have survived very well in that role—too well in fact. But the law of life on a small planet has finally caught up with us, and the standard sentence has been passed.

The glittering array of industrial and medical technology in which we take such pride and on which we have come to depend, is proving to be merely a cunning extension of evolution's plague machinery, providing the very mechanism that locks us into the trap. Lower infant mortality and better public health means more people needing to be fed; higher standards of living and better education means increased consumption of goods and resources and higher levels of waste and pollution. As standards of living improve and birthrates begin to decline, people buy more cars, houses, and television sets and consume more energy and raw materials to build and run them. By this means they multiply their impact on the biosphere to the point that they more than nullify any advantages gained by the declining birthrates. As a result, even though our global fertility rate may gradually decrease, our energy consumption will escalate as before, ensuring that our total impact will continue to fit the standard plague graph with a precision that lies well beyond the realms of mere coincidence.

More than one-third of the world's people are crammed into India and China alone. India is about to launch a massive car manufacturing industry aimed at the 200 million prospective buyers among its middle classes, and China has begun to follow suit. Some 80% of Chinese families already possess a television set, and home shopping programs are very popular. Few can yet afford to buy what they see, but as their affluence increases and the demand for products grows, the environmental impact of their consumerism promises to far outweigh the effects of family planning, better education, and the spread of sexually transmitted diseases. Already deforested, overpopulated, and blanketed with polluted air, these two nations now face social and environmental problems that dwarf those confronting the Western world. The only sources of energy capable of delivering the vast quantity of cheap electricity required to power their cultural leap into the twenty-first century are sulfur-rich coals, hydroelectric energy, and nuclear power. Nuclear power stations are expensive to build and maintain, have a short life and a consequently small harvest factor (the ratio between the total value of the energy and materials outlayed and the value of the power produced during the entire life of the installation). By comparison, nonnuclear power stations are cheap to build, economical to run, and have a long life; so they are the natural choice. But both water power and coal power entail disastrous environmental consequences.

A Dam Shame

It has always been easy to sell the concept of hydroelectricity to governments and their electors on the grounds that above all it delivers "clean" energy, with no embarrassing by-products—other than during construction. Once the plant is fully operational, running costs are minimal and it appears to contribute almost nothing to greenhouse emissions. In public and political mythology it is a zero-emission technology.

If only these fantasies were true. When a dam is built and a valley floods, the decay of the drowned vegetation and the accumulation of methane-producing mud on the valley floor can have at least the same greenhouse impact as a coal-burning power station of similar capacity. Philip Fearnside, an ecologist at Brazil's National Research Institute for Amazonia, analyzed the carbon dioxide and methane emissions from Brazil's Balbina Reservoir and found that it was delivering roughly sixteen times the greenhouse impact of a fossil-fuel power station of similar capacity. What was worse, he said, in the first year of operation (1987) it would have produced about 10 million metric tons of carbon dioxide and 150,000 metric tons of methane—roughly four times the emissions he

Kow Swamp, Murray Valley, Australia. Hydroelectric or irrigation reservoirs that drown whole forests, like this one at Kow Swamp in northern Victoria, generate huge quantities of methane.

measured in 1996. While Fearnside conceded that the reservoir's emissions of carbon dioxide and methane would continue to diminish over time, he maintained that since a large tree trunk can take up to five hundred years to decay in anoxic (oxygen-free) water, the Balbina Reservoir would continue to be more environmentally expensive than a fossil fuel station with similar capacity for at least fifty years, and perhaps indefinitely.[6]

Not all reservoirs are that damaging, however, and the volume of their greenhouse emissions depends to a great extent on the nature of the site. (The thickly vegetated Balbina Valley is so shallow that most of the reservoir is less than 4 meters deep, and the power station has failed to produce the volume of electricity originally proposed.) Latitude and temperature, on the other hand, have little or no effect on the outcome. A research team led by John Rudd from Canada's Freshwater Research Institute in Winnipeg similarly investigated several hydroelectricity sites in northern Manitoba and found large amounts of methane dissolved in the surface layers of the water. At one site the team recorded 7 grams of methane for every square meter of lake surface, while at another site—a flooded peat bog—they found up to 30 grams of methane and between 450 and 1,800 grams of carbon dioxide per square meter. In a similar study conducted at the big Cedar Lake reservoir (1,200 square kilometers [460 square miles]), also in northern Manitoba, Rudd and his team found greenhouse emissions that were almost identical to those from a comparable coal-fired power station.[7]

Once again technology provides no solutions. No matter how clean a technology might appear, energy is inherently expensive. The more we take, the more we must pay, one way or another. Japan offers another spectacular example of this rule. Beset by smog problems and urged on by the construction industry, the Japanese government has consistently chosen the supposedly clean alternative, hydroelectricity. In the past four decades alone Japan has built 1,000 dams and has another 500 on the drawing boards. Water-filtering wetlands have disappeared, and floods no longer flush the river systems. As cheap electricity became available everywhere, industry blossomed in its wake. Consequently, all of Japan's rivers and most of the nation's coastal waters are now thoroughly polluted.[8] Worse, in view of Fearnside and Rudd's work, is that Japan's greenhouse emissions are probably no less than they would have been had they powered their industrial explosion by fossil fuels. Moreover, the massive volume of cement required to build the dams and hydroelectricity plants represents yet another major injection of carbon into the atmosphere.

Cement production releases carbon dioxide in two ways: as part of the chemical reaction that occurs during the conversion of calcium carbonate to calcium oxide in the kilns and through the burning of large quantities of fossil fuels to heat the kilns to the 1450°C required for the chemical reaction to take place. Cement manufacturers now produce 7% of the world's annual carbon dioxide emissions, and global production figures show a growth rate of 5% a year. Cement production is still well behind power generation and vehicle exhausts as a contributor to global warming, but it exceeds the contribution made by all the world's aircraft.[9]

Again technology merely lures us deeper into the environmental trap. Meanwhile our myth-based technoculture keeps us thoroughly bedazzled, entertained, and unable to comprehend the magnitude of our blunder until all the exits are blocked and the consequences are unavoidable. The denouement of our 2-million-year play will not dawn on us until very late, however. We will have to wait until climatic disorder, rising sea levels, rampant famine, social disintegration, and a growing list of pandemics finally bring the human plague to a halt for the full gravity of our predicament to sink in. Nevertheless the truth is creeping up on us even now in a million microscopic forms.

The Gene Traders

Public medical systems are already beginning to feel the impact of drug-resistant strains of tuberculosis, cholera, malaria, and sexually transmitted diseases. Hospitals are being overrun by untreatable "hospital diseases" such as methicillin-resistant *Staphylococcus aureus* (MRSA) and a dangerous new strain of colonic bacteria known as vancomycin-resistant enterococcus (VRE). Their advance is presently unstoppable. Strain number 16 of MRSA is now immune to all but one group of antibiotics (the vancomycin group), and VRE has already developed total immunity to all known antibiotics. Should VRE transmit its vancomycin resistance to MRSA, the world's hospitals would be forced to close, one after the other, as they become infected.

While predictions like this might appear to be unduly alarmist to most people, microbiologists have, in recent times, learned to view bacteria in an entirely new light, for it is only in the past decade or two that their real place in the evolutionary scheme of things has begun to emerge. As outlined in chapters 5 and 6, very near its evolutionary base earth's life appears to have diverged into two streams, bacteria and archaea, or more properly, eubacteria and archaebacteria. This primary division is still visible. Our own body cells, for example, evolved from a remixing of those two streams. In other words, we are composite organisms almost entirely composed of cells that have both eubacterial and archaebacterial components. Of the two, the primary movers and shakers of the planet are the eubacteria. They are spectacularly entrepreneurial in their interaction with the environment as a whole and entirely promiscuous in their genetic negotiations with each other. According to Lynn Margulis and Dorion Sagan, "[eu]bacteria trade genes more frantically than a pit full of commodity traders on the floor of the Chicago Mercantile Exchange."[10] They achieve this continual genetic intercourse in various ways: sometimes by shedding spare copies (plasmids) of some of their genes directly into their fluid environment and taking in other plasmids in return, and sometimes by establishing brief physical links with other bacteria, even other types of bacteria, and directly disseminating their genetic material via these links. The flamboyant genetic promiscuity inherent in this bacterial sex act, known as conjugation, lies at the core of all life and provides the underlying drive for all evolution.[11]

As Margulis and Sagan point out, the earth is essentially a bacterial planet, with bacteria either driving or providing the crucial components of every major biological cycle on earth. It also means that when one bacterium develops immunity to a threat such as a pesticide or an antibiotic, then it is only a matter of time before that bacterial survivor passes on its immunity—not only to others of its kind but also to other types of bacteria that inhabit the same environment and face the same environmental threats. This is the reason some chemical pesticides become ineffective within five years of their successful trials. In other words, it is not a matter of *whether* pathogenic organisms like MRSA will crack the code for total drug resistance, but *when*.

The other factor in their favor is the speed of the bacterial reproductive cycle. Most species run through about fifty generations every twenty-four hours. Aided by a plague of highly mobile human vectors, and also by jet engines that can carry those vectors halfway around the world in less than a day, disease bacteria are about to enter a new golden age. And when that happens, the biosphere's final solution to the human problem, population collapse, can begin in earnest.

Plagues may be nothing new on this fertile planet, but in one respect at least, ours has no precedent. For the very first time the final act of this ancient drama will include a David and Goliath duel between the fast and flaky processes of so-called cultural evolution and the grim juggernaut of biospheric retribution. Darwin versus Lamarck, the tortoise and the hare. And since *Homo sapiens* is not the arcane progeny of some divine Intangible but rather the honest product of painstaking Darwinian evolution, there is no reason to suspect that evolution has launched a monster that is beyond the reach of its normal control mechanisms. Our smooth plague graph and the very typical environmental backlash we have already engendered eloquently testify that those control mechanisms are firmly in place. Having not yet gone cosmically feral or escaped the rigid laws of earthly existence by any other means, we must therefore expect to pay for our extravagant lifestyle at the going earth rate.

The Act-of-God Clause

All the biospheric indicators currently suggest that the prattling prodigy *Homo sapiens* will be no match for the well-oiled machinery of the planetary mousetrap. This Gaian mechanism has dealt very successfully with many plagues over the past 4 billion years, and although this confrontation might be on a larger scale than most, our species is not biologically exceptional when viewed against the dazzling spectrum of earthly life. Yet it does seem as though evolution has taken out a little extra "insurance" in our case. By blindfolding us with racial, political, and religious dogma, our X-factor has provided the biosphere with a very effective backup cover—and an "act-of-God" waiver. And if the grim histories of Malta and Easter Island are anything to go by (see chapter 6), this evolutionary escape clause should come into play toward the middle of the twenty-first century as environmental damage mounts, pandemics begin, and crops begin to fail on a regular basis.

We cannot even rely on the dissemination of better information to improve our chances of evading this genetic *Götterdämmerung*. If the facts don't fit our particular genetic imperatives, we will simply alter them, perceiving only what our genetic imperatives require us to perceive. During the sixteenth and seventeenth centuries in Europe, tens of thousands of witches were tortured and executed on the strength of rock-solid eyewitness accounts of their necromancy. Similarly, for the past 4,000 years many thousands of otherwise respectable, reliable citizens have reported seeing gods, devils, angels, and ghosts; have experienced reincarnation and astral travel; and have been abducted and physically examined—sexually in most cases—by intelligent beings from other planets.

Alien abductions are especially interesting. They represent the modern expression of a hallowed tradition that stretches back to the very beginnings of human history. The abductors used to be devils, monsters, sorcerers, and sundry other evil spirits, and only rarely did they attempt to conceal their nefarious intentions or disguise themselves as angels, swans, or white horses. By contrast, modern aliens have shrewdly chosen to conform to the physiological characteristics most favored by the contemporary culture of the abductee. Consequently, UFO crews who were once tall, blond, and Nordic in appearance now conform to latest Hollywood trends and tend to be smallish, humanoid creatures with overgrown heads and huge childlike eyes. (Large, liquid eyes always did well at the box office.)

The Cult Boom

The more mystic, grandiose, and frankly unbelievable the scenario, the more attractive it is to our mystery-starved hunter-gatherer genes. In those nations where extended family structures have virtually disappeared and genetic relatedness no longer provides a practical glue to cement communities together, any tribelike cults that dispense bizarre mystical dogma on a wholesale basis tend to prosper and proliferate. Only the more dangerous of them ever make headlines, and then only if their devotees commit suicide en masse or their leaders arrange their group departure from this life. Recent examples of such extreme behavior occurred at the People's Temple in Jonestown, Guyana (917 deaths), at the Branch Davidian headquarters in Waco, Texas (74 deaths), and at various premises in the United States and Europe run by the Temple of the Sun (more than 70 deaths). One of Japan's 180,000 registered cults, the Aum

Shinrikyo, or Supreme Truth, was quietly manufacturing a sarin-type World War II nerve gas with the intention of holding the entire Japanese nation, and ultimately the world, for ransom. The Aum factory possessed the capacity to generate enough sarin gas to exterminate the whole human population several times over. A trial attack on a Tokyo subway in 1995 killed only 12 people but severely affected more than 6,000, several thousand of whom were hospitalized. These cults achieve their astonishing hold over their devotees principally because of their rigid tribalism and the patent incredibility of the main tenets of their dogmatic faiths.[12]

Just as dangerous are the secretive networks of hate-filled political and religious fundamentalists that now fester in many regions of the world—Afghanistan, Algeria, the Americas, the Middle East, and Northern Ireland, to name but a few. The membership requirements of such idealistic organizations are demanding indeed. Inductees must be capable of believing that a marketplace littered with the bomb-torn bodies of women and children not only advances the cause of justice and freedom but also entitles the bomber to a place of honor in his chosen afterlife should he die in the process.

Cults proliferate most readily when an idealistic national culture begins to fragment, suggesting that the total volume of mysticism inherent in a culture remains essentially constant. Japan, for example, has given birth to most of its cults and religious sects since the 1960s, whereas the United States and Russia have become breeding grounds for wacky mysticism of all kinds largely since the end of the Cold War. In the same way that many people visit psychoanalysts to solve business and personal problems, so urban Russians now go to their local sorcerers, or *kolduns*. In Moscow alone there are some 15,000 fully licensed sorcerers, all of them registered with the Ministry of Health, and two of Moscow's most popular television programs are entirely devoted to witchcraft and presented by *kolduns*. According to Moscow-based journalist Juliet Butler, "Not since the Middle Ages has witchcraft so possessed the minds of a nation."[13]

Genetic Blinders

In matters mystical the range of human gullibility is practically limitless. And although most of us feel no urge to undergo alien abduction or to attack our local subway station with sarin gas, the widespread acceptance of creationism, astrology, and sustainable economic growth serves as a warning that the fraction of the population capable of applying even the most basic rules of evidence to emotionally derived information is so small as to be inconsequential. The more seductive and dangerous mythologies—those that underpin racism, religion, and politics—are genetically armor-plated and in a rational sense totally bulletproof.

This widespread willingness to suspend disbelief ensures that those who covet tribal leadership and work their way into positions of power will remain blissfully unaware of the quiet avalanche of scientific data that signals biospheric distress in their electorates. It ensures that presidents, prime ministers, and others who wield political power will remain deaf and blind to the only budgetary figures that really matter—the figures that betray our massive environmental debt. Similarly, those responsible for directing the flow of industry and resources will continue to overlook the collateral damage represented by the hemorrhagic loss of species—so long as those species fail to vote. Even Milton Friedman, Nobel laureate in economics, and Julian Simon, professor of business practice at the University of Maryland, have on many occasions stated publicly that direct measures of human welfare are the only indicators that really matter. Simon believes that continued population growth increases wealth and raises living standards, and Friedman believes that technology and human ingenuity can solve all (human) problems. Betraying the genetic origin of their selective vision, they consistently avoid confronting the implications inherent in the decade-long per capita decline in the annual fish catch, the even longer decline in per capita grain production, the accelerating loss of productive land due to various forms of degradation, the soaring extinction rate around the globe, and the carbonization of the atmosphere. The very simple message spelled out by those four crucial indicators of biospheric stress threatens the tribal status and personal security of such men so much that their genes are forced to intervene and bypass the impressive analytical faculties that give them their professional respectability.

This spectacular form of selective blindness is especially widespread in the spheres of politics, commerce, and economics, and it offers an eloquent testimonial to the power of the human X-factor. Flooded by the right blend of neurotransmitters, even the most incisive and disciplined mathematical brain may become wholly impervious to inconvenient mathematical facts. Genetically blinded in this fashion, such men are then able to say with absolute conviction that the "supposed scares don't exist," that overpopulation "is the most incredible triumph of humanity," and that the human brain is "the ultimate resource."[14] All the while, the global population continues to grow, per capita food production continues to decline, atmospheric carbon continues to mount, and the circumference of the earth remains uncharitably constant.

A Final Solution

Judging by the past 600 million years of fossil record, the main thrust of evolution—apart from a half dozen major hiccups—has been toward biological diversity. As suggested by Darwin, and corroborated by fieldwork all around the globe, it is diversity that lends earth's biota its stability and resilience. Even James Lovelock's computerized Daisyworld offered clear confirmation of the link between biodiversity and environmental stability. Given the soaring extinction rate due to human overpopulation, the greatest threat to earth's biota is our own species. From a biospheric point of view, then, whatever contributes to our species' collapse is of paramount evolutionary value. Since global industrialization and rampant consumerism currently offer the biosphere its best chance of achieving a final solution to the human plague, what it now needs from us is more of the same behavior. That means a steady procession of wealthy entrepreneurs and corrupt governments in developing nations and a succession of ambitious, business-oriented governments in developed nations. This combination has produced maximum environmental degradation in the past and will continue to do so by ensuring that those old sacred cows—Growth and Progress—continue to wander the corridors of power unchallenged.

It is not always easy to perfect the political formula for environmental disaster. At the 1997 Kyoto conference on greenhouse emissions, Australia was granted a unique concession, allowing it to continue to increase its emissions by 8% a year (even though most other developed nations had agreed to reduce theirs). The Prime Minister and Minister for the Environment then proudly proclaimed that this peculiar license to pollute represented a "victory for the Australian people" and a "triumph for Australian diplomacy."[15] Armed with deliberate ignorance of science and the mystic's ability to reinvent the world according to their own mythologies, politicians everywhere will thus ensure that our species becomes a willing conspirator in its own demise.

As the work of Hans Selye and others shows, there is only one solution to a mammal plague: catastrophic decline. Whether orchestrated by Selye's general adaptive syndrome (GAS) or by resource collapse, it seems that there are no half measures. So having begun to display a decline that is not yet directly related to food shortage, the big question for us is whether we have culturally mutated ourselves into the rodent mold to the point that we might sustain a full GAS collapse, or whether this will be a low-key primate affair, as demonstrated by wild yellow baboons? Given the meager evidence of fertility decline that is currently available, a reasonable expectation might be that our response will lie somewhere between the two extremes. However, as we have already seen, an ominous pattern of overpopulation and autocollapse has indeed characterized recent human history—especially where the cultures were physically isolated and confined, as on Malta and Easter Island—suggesting that our peculiar vulnerability to mystical delusions may well provide a cultural substitute for the unknown biological factors that send mouse plagues into terminal postplague decline.

Despite the currently falling birthrate, our plague has at least another thirty years of growth ahead of it. (Most authorities, including the United Nations, predict at least another forty to fifty years of growth.[16]) Since we have almost reached the limit of our global food resources, and given our accelerating rate of energy and resource consumption, we seem to be well set up for an environmental coup de grâce in the second half of the twenty-first century. To put it another way, we are facing precisely the same conclusion that all mammal plagues eventually face—a hormonally orchestrated autodecline followed by an environmental backlash that cleans up most of the stragglers.

Excalibur

All adaptive features are, in the final analysis, limitations. If they were not, then life could not have proceeded past the point of global colonization. In other words, adaptive features and their inherent limitations are the very stuff of Darwinian selection, the fabric from which evolution's elegant designs are cut. Compared to other primates we were seriously underendowed, except in one respect—our brain. But the close collaboration that eventually developed between human language and our so-called spirituality not only compensated for our physical shortcomings but also became an evolutionary asset of revolutionary consequence. Here was the magic sword, Excalibur, that would eventually turn this disinherited, endangered primate into a supreme survivor, a behavioral prodigy that would one day manage to meddle with the evolutionary process itself.

But there was a terrible penalty attached to this astonishing asset. By selectively preserving the mystics among our hominid ancestors, evolution devised insurance against our overwhelming success. Only our obsessive yearning for significance, spirituality, and the supernatural—in other words, our X-factor—could have blinded us to the dangers of overpopulation and environmental degradation and prevented us from taking sufficient precautions to avoid it. On the other hand, had our enlarging cortex snatched the reins of behavior from our genes and turned us into fully rational animals, we might have been able to develop a more prudent and sustainable domination of the planet, thereby severely limiting biological diversity and shortchanging evolution itself. And that, from a biospheric point of view, may well have represented the least attractive option of all. No truly Gaian system could tolerate it.

Consequently, like so many animals before us, we were fitted with an adaptive device that was both asset and potential liability, as effective in the short term as it was lethal in the long term. It consisted of an outsized brain equipped with a state-of-the-art language facility that was hot-wired, via our hypothalamus, to our genes. Superbly effective though it was in its original environment, like all adaptive devices it also represented an efficient self-destruct mechanism that would be our undoing when the rules for survival changed.

In our defense it can be said that were it not for our obsession with mysticism, we would have been alien indeed on this cosmic Camelot. Surrounded as we are by such obsessive mystics as the territorial pardalote, the genocidal chimpanzee, and the infanticidal jacana, a wholly rational existence would have been unthinkable. Animals cannot help but sing, dance, mate, and fight in obedience to their genetically directed notions of territorial proprietorship and sexuality. And we are no exception.

The chief goal of all life is genetic survival. Since nothing else matters to genetic material, it is worth any sacrifice. So, like millions of other inhabitants of this planet, we will no doubt continue to do our mystic duty according to the roles genetically assigned to us. Evolution doesn't bend its rules for anyone, not even a talkative, mystical primate. The human plague is following its preordained path, and the final outcome is likely to be grim indeed. Although we have no option but to do our genes' bidding, we would do well to try to spare the biosphere's other progeny to whatever extent we can.

There is a kind of bleak satisfaction to be had in stripping away the fantasy of human spirituality and superiority. Yet, without our genetically derived delusions, we would never have come this far. My ancient hypothalamus instinctively recoils at the prospect of their loss, and my Broca's area crackles with cortical outrage. All of our literature, music, art, drama, history, law, and legend has been wholly founded on our genetically engendered misperceptions, and this virtual reality is the only reality we know.

The gaudy heroes, villains, gods, and monsters that strut the tiny stage of our perceptions may be a poor substitute for the majestic surge of biological reality that washes over, through and about us every second of our lives, but caught in the spotlight of our conscious existence, we can only guess at what is really happening in the genetic darkness that whispers all around us, and only rarely can we catch a glimpse of the immensity of its cosmic context. But despite the folly and pain that mysticism breeds, we should dread its disappearance. We, and the whole animal world, utterly depend on that vital streak of genetically engendered insanity. Without it, no dingo would howl nor nightingale sing. Spring, and all of life, would be a silent thing indeed.

So let us recognize human mysticism for what it really is: the rusting Excalibur of our species, an old and vital streak of genetic madness that once rescued our kind from the brink of extinction, took us to the stars, and will run us through with due dispatch when our little play is done. Ultimately, I have no real argument with mysticism, nor even with the fear and ignorance on which it feeds. The frail, the fearful, and the foolish—these are my kind of animals.

NOTES

Preface

1. Richard Dawkins, *The Blind Watchmaker* (1986), p. xv.
2. For those unfamiliar with the Excalibur reference I should explain that it comes from the Arthurian legends of ancient Britain. Arthur was said to have carried an enchanted sword, known by the name Excalibur, that rendered him invincible in battle.

1. A Prattling Prodigy

1. Laura E. Berk, "Why Children Talk to Themselves," *Scientific American*, November 1994, pp. 60–65.
2. Carl Sagan and Ann Druyan, *Shadows of Forgotten Ancestors* (1992), pp. 276–79.
3. Ibid., pp. 273–74. See also Jared Diamond, *The Rise and Fall of the Third Chimpanzee* (1991), pp. 20–26.
4. Anthropocentrism, the idea that leads us to interpret all things in terms of human experience and human values, is founded in the belief that our globally dominant presence is, in itself, convincing evidence that the earth, if not the cosmos, was "designed with us in mind."
5. Sue Savage-Rumbaugh and Roger Lewin, *Kanzi: The Ape at the Brink of the Human Mind* (1994).

2. Turn of the Tide

1. These words were written by documentary screenwriter Ted Perry in 1972. That they are not those used by Suquamish Chief Seattle in 1854, as is generally believed, erodes neither their accuracy nor their pertinence, then or now.

2. Edward O. Wilson, *The Diversity of Life* (1992), p. 272.

3. Tony McMichael, *Planetary Overload* (1993), p. 112. McMichael notes that the total tonnage of the world's 6 billion human beings is surpassed among vertebrates only by the cattle that they maintain as a food source.

4. Ted Trainer, *The Global Crisis* (1995), p. 27.

5. David M. Raup, *Bad Genes or Bad Luck?* (1991), pp. 70–73.

6. Wilson, *The Diversity of Life*, pp. 182, 275, 278.

7. Ibid., p. 280. (Reproduced by permission of Penguin Books Ltd.)

8. Richard Leakey and Roger Lewin, *The Sixth Extinction: Biodiversity and Its Survival* (1995), pp. 234–41.

9. Claude Martin, "The Year the World Caught Fire," *1997 World Wide Fund for Nature Report* (summary), p. 1.

10. Raup, *Bad Genes or Bad Luck?* p. 73.

11. Ibid. See also Douglas H. Erwin, "The Mother of Mass Extinctions," *Scientific American*, July 1996, pp. 56–62.

12. Edward O. Wilson, "Threats to Biodiversity," *Scientific American*, September 1989, p. 65.

13. Lou Bergeron, "Will El Niño Become El Hombre?" *New Scientist*, 20 January 1996, p. 15.

14. Vin Morgan, research scientist, Australian Antarctic Division (pers. comm., 1996).

15. Tony McMichael, *Planetary Overload* (1993), p. 134. Jeff Hecht, "Bahamas Back Theory of Sudden Climate Change," *New Scientist*, 18 December 1993, p. 14. These figures were obtained from corals and sea-carved cliffs in the Bahamas by Conrad Neumann and Paul Hearty of the University of North Carolina.

16. Debora MacKenzie, "Polar Meltdown Fulfills Worst Predictions," *New Scientist*, 12 August 1995, p. 4; Frank Press and Raymond Siever, *Understanding Earth* (1994), p. 347.

17. Jeff Hecht, "Shallow Methane Could Turn On the Heat," *New Scientist*, 8 July 1995, p. 16.

18. Mopping up radiant heat in a wave band untouched by most other greenhouse gases, methane, mass for mass, is about sixty times more effective than carbon dioxide as a greenhouse forcing agent. Most methane molecules, unlike those of carbon dioxide, persist in the atmosphere for only about ten years, however, and on a hundred-year time frame, methane is only eleven times as potent as carbon dioxide; the factorial average is usually considered to be around twenty.

19. Jeff Hecht, "Baked Alaska," *New Scientist*, 11 October 1997, p. 4.

20. Richard A. Houghton and George M. Woodwell, "Global Climatic Change," *Scientific American*, April 1989, pp. 36–44; John Gribbin, "Methane May Amplify Climate Change," *New Scientist*, 2 June 1990, p. 13.

21. Houghton and Woodwell, "Global Climatic Change," pp. 36–44.

22. Ibid.

23. James Woodford, *Sydney Morning Herald*, 24 June 1995. Also Vin Morgan, research scientist with the Australian Antarctic Division (pers. comm., 1996).

24. Bill de la Mare, "Abrupt Mid-Twentieth-Century Decline in Antarctic Sea-Ice Extent from Whaling Records," *Nature* 389 (1997), p. 57. Between the years 1931 and 1957, and again between 1972 and 1987, British and Norwegian whaling ships customarily hunted blue, humpback, fin, and minke whales at the edge of Antarctica's floating pack ice and daily reported their catches and their positions. When de la Mare plotted each year's position reports on a map of the Antarctic region, he found that they precisely charted the gradual southward retreat of the edge of the pack ice.

25. MacKenzie, "Polar Meltdown," p. 4; Vincent Kiernan, "Is the Frozen North in Hot Water?" *New Scientist*, 8 February 1997, p. 10.

26. Ross Edwards, "Not Yodeling but Drowning," *New Scientist*, 11 November 1995, p. 5.

27. John and Mary Gribbin, "The Greenhouse Effect," *Inside Science: New Scientist*, 13 July 1996, p. 4; David Schneider, "The Rising Seas," *Scientific American*, March 1997, p. 100.

28. Fred Pearce, "Pacific Plankton Go Missing," *New Scientist*, 8 April 1995, p. 5.

29. Barbara E. Brown and John C. Ogden, "Coral Bleaching," *Scientific American*, January 1993, pp. 44–50.

30. Nigel Dudley and Jean-Paul Jeanrenaud, "The Year the World Caught Fire," *1997 World Wide Fund for Nature Report* (summary).

31. Eugene Linden, "Warning from the Ice," *Time Magazine*, 17 March, 1997, pp. 100–105.

32. Paul R. Ehrlich and Anne H. Ehrlich, *Healing the Planet* (1991), pp. 150–51, 203, 210–11.

33. United Nations Environment Program, *Global Environment Outlook* (1997), p. 236.

34. Ehrlich and Ehrlich, *Healing the Planet*, pp. 195–201. See also Lester R. Brown, "Grain Harvest Drops," in *Vital Signs: The Trends That Are Shaping Our Future* (1992), pp. 24–25.

35. McMichael, *Planetary Overload*, p. 207. This figure was originally cited in a 1990 UN report: *Global Outlook 2000. An Economic, Social and Environmental Perspective.* ST/ESA/215/Rev.1.

36. Mary E. White, *Listen . . . Our Land Is Crying* (1997), pp. 85, 91–92.

37. *Australia: State of the Environment 1996*, p. 2:10.

38. *Australia: State of the Environment 1996*, pp. 5:9, 6:12; . Commonwealth Biodiversity Unit, "Native Vegetation Clearance, Habitat Loss and Biodiversity Decline," *Biodiversity Series No. 6* (1995), pp. 14, 18–19.

39. White, *Listen . . .* , p. 107.

40. Fred Pearce, "Thirsty Meals That Suck the World Dry," *New Scientist*, 1 February 1997, p. 7.

41. United Nations Population Fund, *Human Development Report 1994*, p. 6.

42. Fred Pearce, "Poisoned Waters," *New Scientist*, 21 October 1995, pp. 29–33.

43. Ibid.

44. D. E. Gelburd, "Managing Salinity: Lessons from the Past," *Journal of Soil and Water Conservation* 40 (1985), pp. 329–31.

45. McMichael, *Planetary Overload*, pp. 203–8, 216–19. The first ever decline in per capita grain production occurred in 1957, just before the so-called Green Revolution began.

46. Brown, "Grain Harvest Drops," pp. 24–25.

47. Kurt Kleiner, "Panic as Grain Stocks Fall to All Time Low," *New Scientist*, 3 February 1996, p. 10.

48. McMichael, *Planetary Overload*, pp. 216–17. See also Ehrlich and Ehrlich, *Healing the Planet*, p. 197.

49. Vaclav Smil, "Global Population and the Nitrogen Cycle," *New Scientist*, 30 May 1998, p. 63.

50. Joel E. Cohen, *How Many People Can the Earth Support?* (1995), p. 171. See also Ehrlich and Ehrlich, *Healing the Planet*, p. 211.

51. Richard Allison and Ann Greene, "Recombination between Viral RNA and Transgenic Plant Transcripts," *Science* 263 (1994), pp. 1423–25. The extreme motility of genetic material at this level is covered in greater detail in chapter 10.

52. Peter McGrath, "Lethal Hybrid Decimates Harvest," *New Scientist*, 30 August 1997, p. 8.

53. Rob Edwards, "Tomorrow's Bitter Harvest," *New Scientist*, 17 August 1996, p. 14.

54. Rob Edwards, "End of the Germ Line," *New Scientist*, 28 March 1998, p. 22.

55. Bob Holmes, "Blue Revolutionaries," *New Scientist*, 7 December 1996, p. 33.

56. Debora Mackenzie, "The Cod That Disappeared," *New Scientist*, 16 September 1995, pp. 24–29. In April 1997, however, the Canadian government decided to revoke the ten-year fishing ban and allow limited fishing to resume in certain areas.

57. Timothy Flannery, *The Future Eaters*, (1994), pp. 95–96, 102–7; Australian Bureau of Statistics, *Commodity Statistical Bulletin 1994*, p. 107.

58. Holmes, "Blue Revolutionaries," pp. 32–36.

59. Ibid., pp. 33, 36.

60. United Nations Population Fund, *Human Development Report 1994*, p. 27.

61. Cohen, *How Many People?* pp. 209–10. Cohen cites the work of Robert Kates of Brown University, Rhode Island, who arrived at this figure using 2,350 kilocalories (calories) per day as the basic per capita requirement.

62. Wayne Meyer, professor of irrigation at Charles Sturt University, New South Wales (pers. comm., 1997).

63. Fred Pearce, "White Bread Is Green," *New Scientist*, 6 December 1997, p. 10. Research by the same team at Exeter University also showed that the 18,000 megajoules of energy expended in producing, processing, marketing, and delivering the food to each consumer represent almost one-tenth of the national energy budget.

64. United Nations Environment Program, *Global Environment Outlook* (1997), p. 231.

65. McMichael, *Planetary Overload*, p. 108.

66. United Nations Environment Program, *Global Environment Outlook* (1997), p. 231; Fred Pearce, "To Feed the World, Talk to the Farmers," *New Scientist*, 23 November 1996, p. 6.

67. I base this figure on the present (declining) growth rate of about 1.3% per annum and on a personal belief that the population will fail to reach even the lowest UN projection of 7.7 billion by 2050.

68. Ehrlich and Ehrlich, *Healing the Planet*, p. 184; McMichael, *Planetary Overload*, p. 228; Bob Holmes, "Water, water everywhere . . ." *New Scientist*, 17 February 1996, p. 17.

69. McMichael, *Planetary Overload*, p. 112.

70. Fred Pearce, "Northern Exposure," *New Scientist*, 31 May 1997, pp. 24–27.

71. Ibid.

72. M. J. Molina and F. S. Rowland, "Stratospheric Sink for Chloro-Fluoro-Methanes: Chlorine Atom-Catalysed Destruction of Ozone," *Nature* 249 (1974): pp. 810–14.

73. *Australia: State of the Environment 1996*, p. 5:18.

74. "Burning Issue," *New Scientist*, 20 July 1996, p. 13. This ozone decline may have been due in part to a particularly dramatic regional reversal in stratospheric winds that year. However, since such reversals are believed to occur about every twenty-six months, the reduction remains significant.

75. Brenda Dekoker, "An Acid Test," *Scientific American*, October 1995, p. 25.

76. Peter Hadfield, "Raining Acid on Asia," *New Scientist*, 15 February 1997, pp. 16–17.

77. Paul Ehrlich and Anne Ehrlich laid out this equation in *The Population Explosion* (1990), pp. 58–59, and explained it in greater detail in *Healing the Planet*, pp. 7–10.

78. The lowest UN projection for this date is 7.7 billion (*State of the World Population 1997*, p. 4).

79. The UN projection for primary energy consumption in 2050 is 2.6 times the 1990 figure. (United Nations Environment Program, *Global Environment Outlook 1997*, p. 216.)

3. Our Genetic Origins

1. There are several alternative codes for some amino acids, and the stop code may be signaled by any one of three triplets.
2. Richard Dawkins, *The Blind Watchmaker* (1986), pp. 123–26.
3. Richard Dawkins, *The Selfish Gene* (1976), p. 35.
4. Ibid., p. 21.
5. Ibid., p. 36.
6. Chicago Zoological Society news release, 16 August 1996.
7. Dawkins, *Blind Watchmaker*, pp. 169–72.
8. This process is described in elegant detail by evolutionary biologist Stephen Jay Gould in *Hen's Teeth and Horse's Toes* (1983), pp. 177–86.
9. Ian Patterson, "Out of Africa Again . . . and Again," *Scientific American*, April 1997, pp. 46–53.
10. Dean Falk, "Hominid Paleoneurology," *Annual Review of Anthropology*, no. 16 (1987): pp. 13–30. See also Terrence W. Deacon, "The Human Brain," in *The Cambridge Encyclopedia of Human Evolution* (1994), pp. 116–21.
11. Falk, "Hominid Paleoneurology," p. 20.
12. Robert Foley and Robin Dunbar, "Beyond the Bones of Contention," *New Scientist*, 14 October 1989, p. 23. See also Deacon, "Human Brain," pp. 116–17.
13. Falk, "Hominid Paleoneurology," pp. 15–16, 26–27; Nicholas Toth, "The Oldowan Reassessed: A Close Look at Early Stone Artifacts," *Journal of Archaeological Science* 12, no. 2 (March 1985): pp. 101–20; Jean-Pierre Changeux, *Neuronal Man* (1986), p. 236. See also Deacon, "Human Brain," p. 116.
14. Michael C. Corballis, *The Lopsided Ape* (1991), pp. 98–99; Richard F. Thompson, *The Brain* (1985), pp. 31–32. See also Deacon, "Human Brain," p. 116.
15. Thompson, *The Brain*, p. 249.
16. Carla J. Shatz, "The Developing Brain," *Scientific American*, September 1992, p. 38.
17. Thompson, *The Brain*, p. 3.
18. Changeux, *Neuronal Man*, pp. 246–49; Corballis, *Lopsided Ape*, pp. 122–25.
19. Changeux, *Neuronal Man*, p. 249.
20. David R. Shaffer, *Developmental Psychology* (1993), p. 375.
21. Yves Coppens, "East Side Story: The Origin of Humankind," *Scientific American*, May 1994, p. 67; see also L. A. Frakes, *Climates throughout Geologic Time* (1979), p. 235.
22. Foley and Dunbar, "Beyond the Bones of Contention," pp. 24–25.
23. Katharine Milton, "Distribution Patterns of Tropical Plant Foods as an Evolutionary Stimulus to Primate Mental Development," *American Anthropologist* 83, no. 3 (September 1981): pp. 534–48. See also Katharine Milton, "Foraging Behavior and the Evolution of Primate Intelligence," in *Machiavellian Intelligence: Social Expertise and the Evolution of Intellect in Monkeys, Apes, and Humans*, ed. Richard Byrne and Andrew Whiten (New York: Oxford University Press, 1988).
24. Deacon, "Human Brain," p. 116; Changeux, *Neuronal Man*, p. 263.
25. Milton, "Distribution Patterns of Tropical Plant Foods," pp. 534–48.
26. Katharine Milton, "Diet and Primate Evolution," *Scientific American*, August 1993, pp. 70–77.
27. Thompson, *The Brain*, p. 46.
28. Corballis, *The Lopsided Ape*, pp. 69–70.
29. Foley and Dunbar, "Bones of Contention," p. 24. See also Christopher Dean, "Jaws and Teeth," *The Cambridge Encyclopedia of Human Evolution* (1994), p. 57.

30. The phenomenon of paedomorphosis is covered in greater detail by Stephen Jay Gould in several of his essays on evolution, notably "A Biological Homage to Mickey Mouse," in *The Panda's Thumb* (1980), pp. 81–91.

31. Jeffrey T. Laitman, "The Anatomy of Human Speech," *Natural History*, August 1984, pp. 20–27. See also Philip Lieberman, "Human Speech and Language," in *The Cambridge Encyclopedia of Human Evolution* (1994), pp. 134–37.

32. Falk, "Hominid Paleoneurology," p. 26. See also Corballis, *Lopsided Ape*, pp. 307–8.

33. S. A. Barnett, *The Rat* (1963), pp. 12–13.

34. Carl Sagan and Ann Druyan, *Shadows of Forgotten Ancestors* (1992), p. 463 n6.

35. Amy Davis Mozdy, "Pay Attention Rover," *New Scientist*, 10 May 1997, pp. 30–33.

36. Roger Lewin, "Human Origins: The Challenge of Java's Skulls," *New Scientist*, 7 May 1994, pp. 36–40. See also Rick Gore, "The Dawn of Humans," *National Geographic*, May 1997, pp. 96–100.

37. Roger Lewin, *The Origin of Modern Humans* (1993), pp. 1–3.

38. Ofer Bar-Yosef and Bernard Vandermeersch, "Modern Humans in the Levant," *Scientific American*, April 1993, pp. 64–70.

39. Alan Thorne, Australian National University, Canberra (pers. comm., 1996).

40. Ibid.

41. Ibid.

42. The island of New Guinea represents the northern edge of the Australian continental raft, and during periods of low sea level, such as 160,000 to 142,000 years ago and 20,000 to 15,000 years ago, it was linked to mainland Australia by broad, fertile plains that now lie beneath the Arafura Sea and the Gulf of Carpentaria.

43. Mary E. White, *After the Greening* (1994), pp. 188–93; L. A. Frakes, *Climates throughout Geological Time* (1979), p. 249. See also G. Singh and E. A. Geissler, "Late Cenozoic History of Vegetation, Fire, Lake Levels and Climate at Lake George, New South Wales, Australia," *Philosophical Transactions of the Royal Society of London* 311 (1985): pp. 379–447.

44. Leigh Dayton and James Woodford, "Australia's Date with Destiny," *New Scientist*, 7 December 1996, pp. 30–31; A. P. Kershaw, P. T. Moss, and S. van der Kaars, "Environmental Change and the Human Occupation of Australia," *Anthropologie* 35, no. 23 (1997): pp. 35–43. See also A. P. Kershaw, "Climatic Change and Aboriginal Burning in North-East Australia during the Last Two Glacials," *Nature* 322 (1986): pp. 47–49.

45. D. R. Harris, "Human Diet and Subsistence," *The Cambridge Encyclopedia of Human Evolution* (1994), p. 72; Richard E. Leakey and Roger Lewin, *Origins* (1982), p. 123.

46. M. J. Morewood et al., "Fission-Track Ages of Stone Tools and Fossils on the East Indonesian Island of Flores," *Nature* 392, no. 12 (1998): p. 173.

47. Patricia Vickers-Rich and Thomas Hewitt Rich, *Wildlife of Gondwana* (1993), p. 197.

48. Alan Thorne, Australian National University (pers. comm., 1997).

4. The Agrarian Transition

1. Jared Diamond, *The Rise and Fall of the Third Chimpanzee* (1991), p. 168.

2. A more detailed account of this and other examples of monkey culture in action is given by Carl Sagan and Ann Druyan in *Shadows of Forgotten Ancestors* (1992), pp. 348–51. See also Richard Wrangham et al., eds., *Chimpanzee Cultures* (Cambridge: Harvard University Press, 1994).

3. Thomas Robert Malthus, *An Essay on the Principle of Population* (1798).

4. Thomas Malthus's phrase "struggle for existence" appears to have led both Charles Darwin and Alfred Wallace independently to pursue the idea of evolution through natural selection.

5. Originally published in a Danish-Norwegian economic magazine (*Danmarks og Norges Oeconomiske Magazin*) in 1758.

6. Joel E. Cohen, *How Many People Can the Earth Support?* (1995), pp. 77, 400.

7. These figures reflect a sharp fertility decline. According to a 1997 UN Population Fund report, the 1990–1995 growth rate was 1.48% (*The State of World Population 1997*, p. 4). The current growth rate is only 1.33% to 1.35%, and it's falling fast (for more about this, see chapter 6).

8. Cohen, *How Many People?* pp. 5–6, 400.

9. D. E. Gelburd, "Managing Salinity: Lessons from the Past," *Journal of Soil and Water Conservation* 40 (1985): pp. 329–31; Curtis N. Runnels, "Environmental Degradation in Ancient Greece," *Scientific American*, March 1995, pp. 72–75.

10. Andrew Dobson, "People and Disease," *The Cambridge Encyclopedia of Human Evolution* (1994), pp. 415–16.

5. Evolution's Answer to Biological Waste

1. Stephen Jay Gould, *Life's Grandeur* (1996), pp. 221–23.

2. Mike Tyler, zoologist and frog specialist at the Adelaide University, assures me that the boiled frog story is a myth and that frogs are not so long-suffering.

3. Jim Downey, "The Paperless Office: A Science Fiction Fantasy," *Australian Conservation Foundation Report*, 12 June 1996, pp. 2–3.

4. Debora MacKenzie, "Off to a Dirty Start," *New Scientist*, 20 September 1997, p. 13.

5. Fred Pearce, "Catalyst for Warming," *New Scientist*, 13 June 1998, p. 20.

6. Most evolutionary biologists now agree with Lynn Margulis's proposal that the complex internal structure of the modern eukaryote represents the product of a highly evolved eubacterial-archaeal partnership. Both eubacterial and archaeal genes are found in all eukaryote DNA, but only eubacterial genes reside in mitochondrial DNA—and indeed in the small quantities of genetic material that have now been identified in many other eukaryote cell structures. The theory also raises the possibility that the membrane enclosure of the nucleus may even have evolved, not in response to the threat of external toxicity, but to prevent the cell's bacterial lodgers from feeding on the host's DNA and thereby threatening all participants in the chimeric partnership.

7. Lynn Margulis, University of Massachusetts (pers. comm., 1998).

8. With their foundations deeply buried in sediment and a growth rate between 0.5 mm and 1 mm a year, some Shark Bay stromatolites may be more than one thousand years old.

9. Lynn Margulis and Dorion Sagan, *What Is Life?* (1995), p. 133.

6. Correcting Imbalances

1. James Lovelock, *The Ages of Gaia* (1987), p. 63.

2. Lynn Margulis and Dorion Sagan, *What Is Life?* (1995), p. 28.

3. Lovelock, *Ages of Gaia*, pp. 125, 133–35.

4. Richard A. Houghton and George M. Woodwell, "Global Climate Change," *Scientific American*, April 1989, pp. 36–44.

5. Norman R. Pace, "A Molecular View of Microbial Diversity and the Biosphere," *Science* 276 (2 May 1997): p. 736.

6. Stephanie Pain, "The Intraterrestrials," *New Scientist*, 7 March 1998, pp. 28, 32; Pace, "Microbial Diversity and the Biosphere," pp. 736, 739.

7. Jeff Hecht, "Shallow Methane Could Turn On the Heat," *New Scientist*, 8 July 1995, p. 16; Reg Morrison, *Australia: The Four-Billion-Year Journey of a Continent* (1990), pp. 82–87, 126–34. See also L. A. Frakes, *Climates throughout Geologic Time* (1979), pp. 58–61, 129–33.

8. Houghton and Woodwell, "Global Climate Change," pp. 39–40.

9. Lovelock, *Ages of Gaia*, pp. 45–53.

10. Carl Sagan, *The Demon-Haunted World* (1996), p. 117; Ian Lowe, "Are We Really That Smart?" *New Scientist*, 1 July 1995, p. 47.

11. Michael Archer, Tim Flannery, and Gordon Grigg, *Kangaroo* (1985), p. 34–35; Samuel K. Wasser, "Reproductive Control in Wild Baboons Measured by Fecal Steroids," *Biology of Reproduction* 55 (1996): pp. 393–99; M. R. Soules, "Luteal Dysfunction," in *The Ovary* (1993), pp. 607–27.

12. Dennis Chitty, *Do Lemmings Commit Suicide? Beautiful Hypotheses and Ugly Facts* (1996), pp. 104–11, 200–206; Samuel A. Barnett, *The Rat: A Study in Behaviour* (1963), pp. 134–36; John J. Christian, "The Adreno-Pituitary System and Population Cycles in Mammals," *Journal of Mammalogy* 31 (1950): pp. 247–59.

13. Hans Selye, "A Syndrome Produced by Diverse Nocuous Agents," *Nature* 138 (1936), p. 32, (6:1); John B. Calhoun, "Population Density and Social Pathology," *Scientific American*, February 1962, pp. 139–46. See also "The General Adaptation-Syndrome and Diseases of Adaptation," *Journal of Clinical Endocrinology and Metabolism* 6 (1946): pp. 217–30.

14. Chitty, *Do Lemmings Commit Suicide?* pp. 98–99, 128–29.

15. Gunter Dörner, "Prenatal Stress and Possible Aetiogenetic Factors of Homosexuality in Human Males," *Endokrinologie*, no. 75 (1980): pp. 365–68.

16. Gail Vines, "Some of Our Sperm Are Missing," *New Scientist*, 26 August 1995, p. 23–26; Rachel Carson, *Silent Spring* (Houghton Mifflin, 1962).

17. Vines, "Some of Our Sperm Are Missing," pp. 23–26; Beth Martin and Michael Day, "Fresh Alarm over Threatened Sperm," *New Scientist*, 11 January 1997, p. 5.

18. United Nations, *Global Environment Outlook* (1997), p. 224.

19. Frans de Waal, "Bonobo Sex and Society," *Scientific American*, March 1995, pp. 58–64. Frans de Waal is a research professor at Yerkes Regional Primate Center in Atlanta, Georgia.

20. Frans de Waal, *Good Natured* (1996), p. 154; de Waal, "Bonobo Sex and Society," pp. 60, 63.

21. Ibid., p. 60. The other great apes have been seen to mate face-to-face but only very rarely, and only the bonobo is known to change position midstream merely for the sake of variety and to extend their enjoyment of the act.

22. Ibid., p. 58.

23. Ibid., p. 59.

24. F. Gibson, "The Simplest Forms of Life," in *In the Beginning . . .* (1974), p. 118.

25. United Nations Population Fund, *The State of World Population 1997*, p. 4.

26. Jared Diamond, *The Rise and Fall of the Third Chimpanzee* (1991), pp. 297–301. See also Tony McMichael, *Planetary Overload* (1993), pp. 84–87.

27. Paul Bahn and John Flenley, *Easter Island, Earth Island* (1992).

28. Ibid. p. 203–7. See also Paul Bahn, "Who's a Clever Boy Then?" *New Scientist*, 14 February 1998, pp. 44–45.

29. Bahn and Flenley, *Easter Island*, p. 179.

30. Caroline Malone et al., "The Death Cults of Prehistoric Malta," *Scientific American*, December 1993, pp. 76–83. See also S. Stoddart et al., "Cult in an Island Society: Prehistoric Malta in the Period," *Cambridge Archaeological Journal* 3, no. 1 (April 1993): pp. 3–19.

31. Malone et al., "Death Cults," pp. 76–83.
32. Ibid., p. 83.

7. The Terminators

1. J. C. Fanning, M. J. Tyler, and D. J. C. Shearman, "Converting a Stomach to a Uterus: The Microscopic Structure of the Stomach of the Gastric Brooding Frog *Rheobatrachus silus*," *Gastroenterology* 82, no. 1 (1982), pp. 62–70.
2. Michael J. Tyler, *There's a Frog in My Stomach* (1984), pp. 22–43.
3. Ibid.
4. Kurt Kleiner, "Billion-Dollar Drugs Are Disappearing in the Forest," *New Scientist*, 8 July 1995, p. 5.
5. Jared Diamond, *The Rise and Fall of the Third Chimpanzee* (1991), pp. 287–91; Tim Flannery, *The Future Eaters* (1995), pp. 195–98.
6. Flannery, *Future Eaters*, pp. 235–36.
7. *Australia: State of the Environment 1996*, pp. 4:6, 4:16. Almost half of that damage has been done in the last fifty years.
8. Mary E. White, *Listen . . . Our Land Is Crying* (1997), p. 13.
9. Richard Dawkins, *The Blind Watchmaker* (1986), pp. 125–29.
10. C. D. Rowley, *Outcasts in White Australia* (1970), pp. 24, 34–43, 55–58. Rowley was director of the Social Science Research Council of Australia's Aborigines Project from 1964 to 1967.

8. The Spirit in the Gene

1. Roger Lewin, "Birth of a Toolmaker," *New Scientist*, 11 March 1995, pp. 38–41.
2. Richard F. Thompson, *The Brain* (1985), p. 3. Vertebrate brains generally consume only between 2% and 8% of the body's total oxygen intake. However there are several species with a larger EQ than humans. One is an African fish that hunts by a kind of radar system and another is a squirrel monkey whose brain makes up 5% of its total body mass, giving it the largest EQ in the animal kingdom.
3. David Sandeman, professor of neurobiology at the University of New South Wales (pers. comm., 1998); Thompson, *The Brain*, p. 46.
4. David Sandeman, professor of neurobiology, University of New South Wales (pers. comm., 1998).
5. David Sandeman (pers. comm., 1998). See T. H. Bullock and G. A. Horridge, *Structure and Function in the Nervous Systems of Invertebrates*, vol. 2 (1965), pp. 125–323.
6. Judith Rich Harris and Robert M. Liebert, *The Child* (1984), pp. 144–46.
7. Ibid.
8. T. J. Bouchard et al., "Genetic and Environmental Influences on Vocational Interests Assessed Using Adoptive and Biological Families, and Twins Reared Apart and Together," *Journal of Vocational Behavior* 44, no. 3 (1994): pp. 263–78. See also T. J. Bouchard, "Whenever the Twain Shall Meet," *Sciences-New York* 37, no. 5 (1997): pp. 52–57.
9. Harris and Liebert, *The Child*, p. 67; Jean-Pierre Changeux, *Neuronal Man* (1986), pp. 163–68.
10. Florence Levy et al., "Attention-Deficit Hyperactivity Disorder: A Category or a Continuum? Genetic Analysis of a Large-Scale Twin Study," *American Academy of Child and Adolescent Psychiatry* 36, no. 6 (June 1997): pp. 1–7.
11. Hans Eysenck and Michael Eysenck, *Mindwatching* (1981), p. 108.

12. Semir Zeki, "The Visual Image in Mind and Brain," *Scientific American*, September 1992, p. 47.

13. Ibid., pp. 43–50.

14. Ibid.

15. Charles Darwin, *The Descent of Man and Selection in Relation to Sex* (Random House, 1993), p. 734.

16. R. W. Sperry, "Lateral Specialisation in the Surgically Separated Hemispheres," *Neurosciences: Third Study Program* (1974), pp. 5–19. Sperry was awarded the Nobel Prize in 1981 for his work with split-brain patients.

17. Michael S. Gazzaniga, "The Split Brain Revisited," *Scientific American*, July 1998, pp. 35–39. See also M. J. Tramo et al., "Hemispheric Specialization and Interhemispheric Integration," in *Epilepsy and the Corpus Callosum* (1995).

18. Sally P. Springer and Georg Deutsch, *Left Brain, Right Brain* (1981), p. 192.

19. Anne Moir and David Jessel, *Brain Sex* (1989), pp. 39–49.

20. Thompson, *The Brain*, p. 26.

21. Carl Sagan, *The Demon-Haunted World* (1996), p. 349.

22. Thompson, *The Brain*, pp. 19–21.

23. Gordon M. Shepherd, *Neurobiology*, 3d ed. (1994), pp. 606–8; Larry Cahill et al., *Proceedings of the National Academy of Sciences* 93 (1996): p. 8016.

24. Alison Motluk, "Touched by the Word of God," *New Scientist*, 8 November 1997, p. 7.

25. Laura Betzig, "Roman Polygyny," *Ethology and Sociobiology* 13, (1992), pp. 309–49.

26. Jared Diamond, *The Rise and Fall of the Third Chimpanzee* (1988), p. 75.

27. Carl Sagan and Ann Druyan, *Shadows of Forgotten Ancestors* (1992), p. 329. Sagan and Druyan note that the gibbons are our nearest relatives that display this behavior, and even then it appears in only this one species.

28. Evolution consists of the continual diversification and selection of species best adapted to prevailing environments. Since environments continually change, the term *progress* is inaccurate and inappropriate.

9. Excalibur!

1. Jared Diamond, "What Are Men Good For?" *Natural History*, May 1993, pp. 24–29.

2. Martin Scorsese, in his narration of the television documentary series "Century of Cinema," directed by Scorsese and Michael Henry Wilson.

3. Dean Falk, "Hominid Paleoneurology," *Annual Review, Anthropology*, no. 16 (1987), p. 26.

4. David Harris, "Human Diet and Subsistence," *Cambridge Encyclopaedia of Human Evolution* (1992), p. 72; Richard Leakey, *Origins* (1991), p. 123.

5. Terrence W. Deacon, "The Human Brain," *The Cambridge Encyclopedia of Human Evolution* (1992), p. 116–18. See also Falk, "Hominid Paleoneurology," pp. 18–25.

6. Bill Neidjie, Stephen Davis, and Allan Fox, *Kakadu Man* (1985), pp. 33–95.

7. R. M. Berndt, *Love Songs of Arnhem Land* (1976), p. 103. According to the late professor Berndt, "This small creature is called a mouse by Aborigines." He adds, "However, it is probably the small marsupial rat; the female of the species is pouched." In fact there is no such thing as a pouched rodent, but the region has four carnivorous mouselike marsupials known as dasyurids, as well as several mouselike rodents, but none of these "hop." This coastal region of Arnhemland does have one little-known hopping mouse (*Notomys aquilo*). Though not a marsupial, this seems a very likely subject for these songs. So, for the sake of accuracy and aesthetics, I have substituted the word *mice* in this instance.

8. B. Price, "AIB National Poll of First-Year Biology Students in Australian Universities," *The Skeptic* 12, no. 3 (1992), pp. 26–31.

9. Michael Archer, " 'Sine' of the Times," *Australian Natural History*, Summer 1994/95, pp. 68–69. See also Ian Lowe, "Are We Really That Smart?" *New Scientist*, 1 July 1995, p. 47.

10. David Attenborough, *The Trials of Life* (1990), pp. 231–34.

11. Michael Jordan, *Cults, Prophesies, Practices, and Personalities* (1996), pp. 68–69.

12. Cited in Frans de Waal's book *Good Natured* (1996), p. 195, the figures were originally quoted in the *New York Times*, 5 August 1990.

13. Jared Diamond, *The Rise and Fall of the Third Chimpanzee* (1988), pp. 250–76.

14. Richard Wrangham and Dale Peterson, *Demonic Males* (1996), pp. 5–6, 10–21. Wrangham was a member of Goodall's research team during this period. These and other premeditated attacks, some resulting in death, are recorded in detail by Jane Goodall in "Life and Death at Gombe," *National Geographic*, May 1979, pp. 592–620. See also Jane Goodall, *Through a Window: My Thirty Years with the Chimpanzees of Gombe* (Boston: Houghton Mifflin, 1990).

15. Wrangham and Peterson, *Demonic Males*, pp. 19–21.

16. Diamond, *Third Chimpanzee*, pp. 252–55; Tim Flannery, *The Future Eaters* (1994), pp. 313–15.

17. Henry Reynolds, *Frontier* (1987), p. 38. The quoted words are commonly attributed to Captain Arthur Phillip, Australia's first governor, who ruled the colony from 1788 to 1792.

18. Flannery, *Future Eaters*, p. 317; see also Reynolds, *Frontier*, pp. 3–80.

19. Diamond, *Third Chimpanzee*, pp. 252–54.

20. Words used in 1859 by William Jowitt, a prominent explorer and author, in a letter sent to his mother in England (Reynolds, *Frontier*, p. 42).

21. Diamond, *Third Chimpanzee*, p. 254; Reynolds, *Frontier*, p. 196.

22. Frans de Waal, *Good Natured* (1996), pp. 125–36, 164–66, 176–82.

23. Ibid., p. 161; Wrangham and Peterson, *Demonic Males*, pp. 127–52. Wrangham and Peterson also note that primatologist Biruté Galdikas reported witnessing the rape of a female member of her camp staff by a young male orangutan named Gundul. He was so strong the two women could do nothing to prevent it.

24. Wrangham and Peterson, *Demonic Males*, pp. 132–52.

25. Richard H. Scheller and Richard Axel, "How Genes Control an Innate Behavior," *Scientific American*, March 1984, pp. 44–52.

26. Richard Frederick Thompson, *The Brain* (1985), pp. 18–19, 34–42.

27. Helen E. Fisher, "Lust, Attraction, and Attachment in Mammalian Reproduction," *Human Nature* 9, no. 1 (1998): pp. 23–52.

28. M. Garry et al., "Imagination Inflation: Imagining a Childhood Event Inflates Confidence That It Occurred," *Psychonomic Bulletin and Review* 3, no. 2 (1996): pp. 208–14; Gordon M. Shepherd, *Neurobiology* (1994), pp. 615–16. See also Elizabeth Loftus, "Creating False Memories," *Scientific American*, September 1997, pp. 51–55.

29. Rosie Mestel, "Behind the Mask," *New Scientist*, 27 April 1996, pp. 10–13. See also Michael S. Gazzaniga, "The Split Brain Revisited," *Scientific American*, July 1998, pp. 35–39.

30. J. G. Hildebrand and G. M. Shepherd, "Mechanisms of Olfactory Discrimination: Converging Evidence for Common Principles across Phyla," *Annual Review Neuroscience* 20 (1997), pp. 595–631; David Sandeman, "Funktionelle Ähnlichkeiten zwischen Nervensystemenvon Vertebraten und Invertibraten: Homolog oder Analog?" *Abh. der Akad. Wiss und Lit. Mainz*, Stuttgart, 1996. David Sandeman is a professor of neurobiology at the University of New South Wales, Sydney.

31. Shepherd, *Neurobiology*, pp. 603–14.

32. David R. Bergamini, *Japan's Imperial Conspiracy* (1971), pp. 202–3.

33. Ibid., p. 54. See also John Costello, *The Pacific War* (1981), p. 578.

34. Bergamini, *Japan's Imperial Conspiracy*, pp. 1036, 1040.

35. Ibid., pp. 1016–30.

36. Ibid., p. 1040. See also Costello, *Pacific War*, pp. 558–59.

37. Bergamini, *Japan's Imperial Conspiracy*, pp. 1030–31.

38. Richard Leakey and Roger Lewin, *The Sixth Extinction: Biodiversity and Its Survival* (1995), p. 241.

39. Sue Armstrong, "Female Circumcision: Fighting a Cruel Tradition," *New Scientist,* 2 February 1991, pp. 22–27; Wrangham and Peterson, *Demonic Males*, p. 119. See also Marguerite Holloway, "Trends in Women's Health," *Scientific American,* August 1994, pp. 77–83.

40. Carl Sagan and Ann Druyan, *Shadows of Forgotten Ancestors* (1992), pp. 324–26; S. T. Emlen, N. J. Demong, and D. Emlen, "Experimental Induction of Infanticide in Female Wattled Jacanas," *The Auk* 105 (1989): pp. 1–7.

10. Midnight Prognosis

1. Mark Buchanan, "Fascinating Rhythm," *New Scientist,* 3 January 1998, p. 22.

2. Tony McMichael, *Population Overload* (1993), p. 204; Paul Ehrlich and Anne Ehrlich, *Healing the Planet* (1991), p. 196; Joel E. Cohen, *How Many People Can the Earth Support?* (1995), p. 187.

3. L. A. Frakes, *Climates throughout Geological Time,* (1979) 1980, p. 236.

4. "Lake Gas," *New Scientist,* 16 August 1997, p. 21. The carbon isotopes in methane allow it to be carbon dated.

5. Steve Meacham, "Protective Reflex Made Apocalypse Cafe Heroes," *Sydney Morning Herald,* 3 May 1996, p. 13.

6. Fred Pearce, "Trouble Bubbles for Hydropower," *New Scientist,* 4 May 1996, pp. 28–30.

7. Ibid., pp. 30–31.

8. Fred Pearce, "Land of the Rising Concrete," *New Scientist,* 11 January 1997, p. 43.

9. Fred Pearce, "The Concrete Jungle Overheats," *New Scientist,* 19 July 1997, p. 14.

10. Lynn Margulis and Dorion Sagan, *What Is Life?* (1995), p. 73.

11. Ibid., pp. 73–76.

12. Michael Jordan, *Cults, Prophesies, Practices, and Personalities* (1996), pp. 58–67.

13. Juliet Butler, "Russia's Witches: Spellbinding a Nation," *Marie Claire Australia,* August 1997, pp. 54–60.

14. Professor Julian Simon, as quoted by Anita Gordon and David Suzuki in *It's a Matter of Survival* (1990), p. 172.

15. Australian Broadcasting Corporation news broadcasts, 11 December 1997. These and other similar statements were made both in parliament and during press interviews given by Prime Minister John Howard and Environment Minister Robert Hill.

16. United Nations Population Fund, *The State of the World Population 1997,* p. 249. (The predicted population growth, ranging from 7.7 to 11.1 billion by 2050, is likely to be revised downward yet again in the 1998 report.)

BIBLIOGRAPHY

Allison, Richard, and Ann Greene. "Recombination between Viral RNA and Transgenic Plant Transcripts." *Science* 263 (March 1994).

Archer, Michael. " 'Sine' of the Times." *Australian Natural History* (summer 1994/95).

Archer, Michael, Tim Flannery, and Gordon Grigg. *Kangaroo.* Sydney: Weldons, 1985.

Armstrong, Sue. "Female Circumcision: Fighting a Cruel Tradition." *New Scientist,* 2 February 1991.

Attenborough, David. *The Trials of Life.* London: William Collins, 1990.

Australia. *Landcover Disturbance over the Australian Continent.* Biodiversity Series, Paper No. 7. Canberra: Department of Environment, Sport and Territories, 1995.

———. *Native Vegetation Clearance, Habitat Loss and Biodiversity Decline.* Biodiversity Series, Paper No. 6. Department of Environment, Sport and Territories, 1995.

———. *Australia: State of the Environment 1996.* Melbourne: CSIRO Publishing.

Bahn, Paul. "Who's a Clever Boy Then?" *New Scientist,* 14 February 1998.

Bahn, Paul, and John Flenley. *Easter Island, Earth Island.* London: Thames and Hudson, 1992.

Barnett, S. A. *The Rat.* Chicago: University of Chicago Press, 1975.

Bar-Yoseph, Ofer, and Bernard Vandermeersch. "Modern Humans in the Levant." *Scientific American,* April 1993.

Bergamini, David. *Japan's Imperial Conspiracy.* London: William Heinemann, 1971.

Berk, Laura E. "Why Children Talk to Themselves." *Scientific American,* November 1994.

Berndt, R. M. *Love Songs of Arnhemland.* Melbourne: Thomas Nelson, 1976.

Berndt, Thomas J. *Child Development.* New York: Holt, Rinehart and Winston, 1992.

Betsworth, D. G., T. J. Bouchard, C. R. Cooper, H. D. Grotevant, J. I. C. Hansen, S. Scarr, and R. A. Weinberg. "Genetic and Environmental Influences on Vocational

Interests Assessed Using Adoptive and Biological Families, and Twins Reared Apart and Together." *Journal of Vocational Behavior* 44, no. 3 (1994).

Betzig, Laura. "Roman Polygyny." *Ethology and Sociobiology* 13 (1992).

Blackmore, Susan. "Alien Abduction." *New Scientist*, 19 November 1994.

Blumenschine, Robert J., and John A. Cavallo. "Scavenging and Human Evolution." *Scientific American*, October 1992.

Bouchard, T. J. "Whenever the Twain Shall Meet." *Sciences–New York* 37, no. 5 (1997).

Broecker, Wallace S. "Chaotic Climate." *Scientific American*, November 1995.

Brown, Barbara E., and John C. Ogden. "Coral Bleaching." *Scientific American*, January 1993.

Brown, L. R. "Grain Harvest Drops." In *Vital Signs: The Trends That Are Shaping Our Future*, edited by L. R. Brown, C. Flavin, and H. Kane. New York: Norton, 1992.

Buchanan, Mark. "Fascinating Rhythm." *New Scientist*, 3 January 1998.

Bullock, T. H., and G. A. Horridge. *Structure and Function in the Nervous Systems of Invertebrates*, no. 12. San Francisco: W. H. Freeman, 1965.

Butler, Juliet. "Russia's Witches: Spellbinding a Nation." *Marie Claire Australia*, August 1997.

Cahill, Larry, et al. *Proceedings of the National Academy of Sciences*, no. 93 (1996).

Calhoun, John B. "Population Density and Social Pathology." *Scientific American*, February 1962.

Changeux, Jean-Pierre. *Neuronal Man*. New York: Oxford University Press, 1986.

Chitty, Dennis. *Do Lemmings Commit Suicide? Beautiful Hypotheses and Ugly Facts*. New York: Oxford University Press, 1996.

Christian, John J. "The Adreno-Pituitary System and Population Cycles in Mammals." *Journal of Mammalogy* 31 (1950).

Cohen, Joel E. *How Many People Can the Earth Support?* New York: W. W. Norton, 1995.

Coppens, Yves. "East Side Story: The Origin of Humankind." *Scientific American*, May 1994.

Corballis, Michael C. *The Lopsided Ape*. New York: Oxford University Press, 1991.

Costello, John. *The Pacific War*. New York: Rawson Wade Publishers, 1981.

Darwin, Charles Robert. *On the Origin of Species*. 1859.

———. *The Descent of Man and Selection in Relation to Sex*. 1871.

———. *The Expression of the Emotions in Man and Animals*. 1872.

Dawkins, Richard. *The Selfish Gene*. New York: Oxford University Press, 1976.

———. *The Blind Watchmaker*. London: Longmans, 1986.

Dayton, Leigh, and James Woodford. "Australia's Date with Destiny." *New Scientist*, 7 December 1996.

Deacon, Terrence W. "The Human Brain." In *The Cambridge Encyclopedia of Human Evolution*, edited by Steve Jones, David Martin, and David Pilbeam. Cambridge: Cambridge University Press, 1992.

Dean, Christopher. "Jaws and Teeth." In *The Cambridge Encyclopedia of Human Evolution*, edited by Steve Jones, David Martin, and David Pilbeam. Cambridge: Cambridge University Press, 1992.

Dekoker Brenda. "An Acid Test." *Scientific American*, October 1995.

de la Mare, Bill. "Abrupt Mid-Twentieth-Century Decline in Antarctic Sea-Ice Extent from Whaling Records." *Nature* 389 (1997).

Diamond, Jared. *The Rise and Fall of the Third Chimpanzee*. London: Radius, 1991.

———. "What Are Men Good For?" *Natural History*, May 1993.

Dobson, Andrew. "People and Disease." In *The Cambridge Encyclopedia of Human Evolution*, edited by Steve Jones, David Martin, and David Pilbeam. Cambridge: Cambridge University Press, 1992.

Dörner, Gunter. "Prenatal Stress and Possible Aetiogenetic Factors of Homosexuality in Human Males." *Endokrinologie*, no. 75 (1980).

Downey, Jim. "The Paperless Office: A Science Fiction Fantasy." *Australian Conservation Foundation Report,* 12 June 1996.

Dudley, Nigel, and Jean-Paul Jeanrenaud. "The Year the World Caught Fire." *WWF International Discussion Paper* (executive summary), December 1997.

Edwards, Rob. "Tomorrow's Bitter Harvest." *New Scientist,* 17 August 1996.

———. "End of the Germ Line." *New Scientist,* 28 March 1998.

Edwards, Rob, and Ian Anderson. "Seeds of Wrath." *New Scientist,* 14 February 1998.

Edwards, Ross. "Not Yodelling but Drowning." *New Scientist,* 11 November 1995.

Ehrlich, Paul R. *The Population Bomb.* New York: Ballantine Books, 1968.

Ehrlich, Paul R., and Anne H. Ehrlich. *The Population Explosion.* New York: Simon & Schuster, 1990.

———. *Healing the Planet.* New York: Addison-Wesley, 1991.

Emlen, S. T., N. J. Demong, and D. Emlen, "Experimental Induction of Infanticide in Female Wattled Jacanas." *The Auk,* no. 105 (1989).

Erwin, Douglas H. "The Mother of Mass Extinctions." *Scientific American,* July 1996.

Eysenck, Hans, and Michael Eysenck. *Mindwatching.* London: Michael Joseph, 1981.

Falk, Dean. "Hominid Paleoneurology." *Annual Review of Anthropology,* no. 16 (1987).

Fanning, J. C., M. J. Tyler, and D. J. C. Shearman. "Converting a Stomach to a Uterus: The Microscopic Structure of the Stomach of the Gastric Brooding Frog *Rheobatrachus silus." Gastroenterology* 82, no. 1 (1982).

Fisher, Helen E. "Lust, Attraction, and Attachment in Mammalian Reproduction." *Human Nature* 9, no. 1 (1998).

Flannery, Timothy Fridtjof. *Australia's Vanishing Mammals.* Sydney: Readers Digest, 1990.

———. *The Future Eaters.* Sydney: Reed Books, 1994.

Foley, Robert, and Robin Dunbar. "Beyond the Bones of Contention." *New Scientist,* 14 October 1989.

Frakes, L. A. *Climates throughout Geological Time.* Amsterdam: Elsevier Scientific, 1979.

Garry, M., C. G. Manning, E. F. Loftus, and S. J. Sherman. "Imagination Inflation: Imagining a Childhood Event Inflates Confidence That It Occurred." *Psychonomic Bulletin and Review* 3, no. 2 (1996).

Gazzaniga, Michael S. "The Split Brain Revisited." *Scientific American,* July 1998.

Gelburd, Diane E. "Managing Salinity: Lessons from the Past." *Journal of Soil and Water Conservation* 40, no. 4 (1985).

Gibson, F. "The Simplest Forms of Life." In *In the Beginning. . . .* Canberra: Australian Academy of Science, 1974.

Goodall, Jane. "Life and Death at Gombe." *National Geographic,* May 1979.

Goodall, Jane. *Through a Window: My Thirty Years with the Chimpanzees of Gombe.* Boston: Houghton Mifflin, 1990.

Gordon, Anita, and David Suzuki. *It's a Matter of Survival.* Toronto: Stoddart Publishing, 1990.

Goss, Helen. "Meltdown Warning as Tropical Glaciers Trickle Away." *New Scientist,* 24 June 1995.

Gould, Stephen Jay. *The Panda's Thumb.* New York: W. W. Norton, 1980.

———. *Hen's Teeth and Horse's Toes.* New York: W. W. Norton, 1983.

———. *Life's Grandeur (Full House).* London: Random House, 1996.

Gribbin, John. "Methane May Amplify Climate Change." *New Scientist,* 2 June 1990.

Gribbin, John, and Mary Gribbin. "Inside Science: The Greenhouse Effect." *New Scientist,* 13 July 1996.

Grove, Richard H. "Origins of Western Environmentalism." *Scientific American,* July 1992.

Hadfield, Peter. "Raining Acid on Asia." *New Scientist,* 15 February 1997.

Harris, D. R. "Human Diet and Subsistence." In *The Cambridge Encyclopedia of Human Evolution*, edited by Steve Jones, David Martin, and David Pilbeam. Cambridge: Cambridge University Press, 1992.

Harris, Judith Rich, and Robert M. Liebert. *The Child*. New Jersey: Prentice Hall, 1987.

Hecht, Jeff. "Bahamas Back Theory of Sudden Climate Change." *New Scientist*, 18 December 1993.

———. "Shallow Methane Could Turn On the Heat." *New Scientist*, 8 July 1995.

———. "Baked Alaska." *New Scientist*, 11 October 1997.

Hildebrand, J. G., and G. M. Shepherd. "Mechanisms of Olfactory Discrimination: Converging Evidence for Common Principles across Phyla." *Annual Review of Neuroscience* 20 (1997).

Holloway, Marguerite. "Trends in Women's Health." *Scientific American*, August 1994.

Holmes, Bob. "Blue Revolutionaries." *New Scientist*, 7 December 1996.

Houghton, Richard A., and George M. Woodwell. "Global Climatic Change." *Scientific American*, April 1989.

Jordan, Michael. *Cults, Prophesies, Practices, and Personalities*. Sydney: The Book Company, 1996.

Kazantzakis, Nikos. *Zorba the Greek*. Translated by Carl Wildman. London: Faber and Faber, 1961.

Kershaw, A. P. "Climatic Change and Aboriginal Burning in North-East Australia during the Last Two Glacials." *Nature* 322 (1986).

Kershaw, A. P., P. T. Moss, and S. van der Kaars. "Environmental Change and the Human Occupation of Australia." *Anthropologie* 35, no. 2/3 (1997).

Kiernan, Vincent. "Is the Frozen North in Hot Water?" *New Scientist*, 8 February 1997.

Krebs, Charles J. *Ecology: The Experimental Analysis of Distribution and Abundance*. New York: Harper and Row, 1972.

Laitman, Jeffery T. "The Anatomy of Human Speech." *Natural History*, August 1984.

Leakey, Richard, and Roger Lewin. *Origins*. London: Macdonald and Jane's Publishers, 1977.

Leakey, Richard, and Roger Lewin. *The Sixth Extinction: Biodiversity and Its Survival*. New York: Doubleday, 1995.

Levy, Florence, David A. Hay, Michael McStephen, Catherine Wood, and Irwin Waldman. "Attention-Deficit Hyperactivity Disorder: A Category or a Continuum? Genetic Analysis of a Large-Scale Twin Study." *American Academy of Child and Adolescent Psychiatry* 36, no. 6 (1997).

Lewin, Roger. *The Origin of Modern Humans*. New York: W. H. Freeman, 1993.

———. "Human Origins: The Challenge of Java's Skulls." *New Scientist*, 7 May 1994.

Lieberman, Philip. "Human Speech and Language." *The Cambridge Encyclopedia of Human Evolution*, edited by Steve Jones, David Martin, and David Pilbeam. Cambridge: Cambridge University Press, 1992.

Linden, Eugene. "Warning from the Ice." *Time Magazine*, 17 March 1997.

Loftus, Elizabeth. "Creating False Memories." *Scientific American*, September 1997.

Lovelock, James. *The Ages of Gaia*. New York: Oxford University Press, 1987.

Lowe, Ian. "Are We Really That Smart?" *New Scientist*, 1 July 1995.

MacKenzie, Debora. "Will Tomorrow's Children Starve?" *New Scientist*, 3 September 1994.

———. "Polar Meltdown Fulfils Worst Predictions." *New Scientist*, 12 August 1995.

———. "The Cod That Disappeared." *New Scientist*, 16 September 1995.

———. "Off to a Dirty Start." *New Scientist*, 20 September 1997.

Malone, Caroline, Anthony Bonanno, Tancred Gouder, Simon Stoddart, and David Trump. "The Death Cults of Prehistoric Malta." *Scientific American*, December 1993.

Malthus, Thomas Robert. *An Essay on the Principle of Population*. 1798.

Margulis, Lynn, and Dorion Sagan. *What Is Life?* London: Weidenfeld and Nicolson, 1995.

Martin, Beth, and Michael Day. "Fresh Alarm over Threatened Sperm." *New Scientist,* 11 January 1997.

McGrath, Peter. "Lethal Hybrid Decimates Harvest." *New Scientist,* 30 August 1997.

McMichael, Tony. *Planetary Overload.* Cambridge: Cambridge University Press, 1993.

Mestel, Rosie. "Behind the Mask." *New Scientist,* 27 April 1996.

Milton, Katharine. "Distribution Patterns of Tropical Plant Foods as an Evolutionary Stimulus to Primate Mental Development." *American Anthropologist* 83, no. 3 (1981).

———. "Diet and Primate Evolution." *Scientific American,* August 1993.

Moir, Anne, and David Jessel. *Brain Sex.* New York: Bantam, Doubleday, Dell, 1992.

Morewood, M. J., P. O. Sullivan, F. Aziz, and A. Raza. "Fission-Track Ages of Stone Tools and Fossils on the East Indonesian Island of Flores." *Nature* 392, no. 12 (1998).

Morrison, Reg. *Australia: The Four-Billion-Year Journey of a Continent.* New York: Facts on File, 1990.

Motluk, Alison. "Touched by the Word of God." *New Scientist,* 8 November 1997.

Mozdy, Amy Davis. "Pay Attention Rover." *New Scientist,* 10 May 1997.

Neidjie, Bill, Stephen Davis, and Allan Fox. *Kakadu Man.* Sydney: Allan Fox and Associates, 1985.

Pace, Norman R. "A Molecular View of Microbial Diversity and the Biosphere." *Science,* 2 May 1997.

Pain, Stephanie. "Tiny Killers Bloom in Warmer Seas." *New Scientist,* 24 February 1996.

———. "The Intraterrestrials." *New Scientist,* 7 March 1998.

Patterson, Francine, and Eugene Linden. *The Education of Koko.* New York: Holt, Rinehart and Winston, 1981.

Patterson, Ian. "Out of Africa Again . . . and Again." *Scientific American,* April 1997.

Pearce, Fred. "How Disappearing Lakes Are Swelling the Oceans." *New Scientist,* 22 January 1994.

———. "Forests Destined to End in the Mire." *New Scientist,* 7 May 1994.

———. "Will Global Warming Plunge Europe into an Ice Age?" *New Scientist,* 19 November 1994.

———. "Dead in the Water." *New Scientist,* 4 February 1995.

———. "Pacific Plankton Go Missing." *New Scientist,* 8 April 1995.

———. "Global Alert Over Malaria." *New Scientist,* 13 May 1995.

———. "Poisoned Waters." *New Scientist,* 21 October 1995.

———. "How the Soviet Seas Were Lost." *New Scientist,* 11 November, 1995.

———. "Trouble Bubbles for Hydropower." *New Scientist,* 4 May 1996.

———. "To Feed the World, Talk to the Farmers." *New Scientist,* 23 November 1996.

———. "Land of the Rising Concrete." *New Scientist,* 11 January 1997.

———. "Thirsty Meals That Suck the World Dry." *New Scientist,* 1 February 1997.

———. "Northern Exposure." *New Scientist,* 31 May 1997.

———. "The Concrete Jungle Overheats." *New Scientist,* 19 July 1997.

———. "Promising the Earth." *New Scientist.,* 30 August 1997.

———. "White Bread is Green." *New Scientist,* 6 December 1997.

Potts, Richard. "The Hominid Way of Life." *The Cambridge Encyclopedia of Human Evolution,* edited by Steve Jones, David Martin, and David Pilbeam. Cambridge: Cambridge University Press, 1992.

Press, Frank, and Raymond Siever. *Understanding Earth.* New York: W. H. Freeman, 1994.

Price, B. "AIB National Poll of First-Year Biology Students in Australian Universities." *The Skeptic* 12, no. 3 (1992).

Rahmstorf, Stefan. "Ice-Cold in Paris." *New Scientist,* 8 February 1997.

Ramage, Colin S. "El Niño." *Scientific American,* June 1986.

Raup, David M. *Bad Genes or Bad Luck?* New York: W. W. Norton, 1991.

Reynolds, Henry. *Frontier.* Sydney: Allen and Unwin, 1987.

Rowley, C. D. *Outcasts in White Australia.* Canberra: Australian National University Press, 1970.

Runnels, Curtis N. "Environmental Degradation in Ancient Greece." *Scientific American,* March 1995.

Safina, Carl. "Where Have All the Fishes Gone?" *Issues in Science and Technology* 10 (spring 1994).

———. "The World's Imperiled Fish." *Scientific American,* November 1997.

Sagan, Carl, and Ann Druyan. *Shadows of Forgotten Ancestors.* London: Random House, 1992.

———. *Demon Haunted World.* London: Headline Book Publishing, 1996.

Sandeman, David. "Funktionelle Ähnlichkeiten zwischen Nervensystemenvon Vertebraten und Invertibraten: Homolog oder Analog?" *Abh. der Akad. Wiss und Lit. Mainz,* Stuttgart. 1996.

Savage-Rumbaugh, Sue, and Roger Lewin. *Kanzi: The Ape at the Brink of the Human Mind.* New York: Wiley, 1994.

Scheller, Richard H., and Richard Axel. "How Genes Control an Innate Behavior." *Scientific American,* March 1984.

Schneider, David. "The Rising Seas." *Scientific American,* March 1997.

Selye, Hans. "A Syndrome Produced by Diverse Nocuous Agents." *Nature* 138, no. 32 (1936).

———. "The General Adaptation-Syndrome and Diseases of Adaptation." *Journal of Clinical Endocrinology and Metabolism,* no. 6 (1946.)

Shaffer, David R. *Developmental Psychology.* California: Brooks/Cole Publishing, 1992.

Shatz, Carla J. "The Developing Brain." *Scientific American,* September 1992.

Shepherd, Gordon M. *Neurobiology.* 3d ed. New York: Oxford University Press, 1994.

Singh, G., and E. A. Geissler. "Late Cenozoic History of Vegetation, Fire, Lake Levels, and Climate at Lake George, New South Wales, Australia." *Philosophical Transactions of the Royal Society of London,* no. 311 (1985).

Smil, Vaclav. "Global Population and the Nitrogen Cycle." *Scientific American,* July 1997.

Soules, M. R. "Luteal Dysfunction." In *The Ovary,* ed. Eli Y. Adashi and Peter C. K. Leung. New York: Raven Press, 1993.

Sperry, R W. "Lateral Specialization in the Surgically Separated Hemispheres." In *Neurosciences: Third Study Program,* ed. F. O. Schmitt and F. G. Worden. Cambridge: MIT Press, 1974.

Springer, Sally, and Georg Deutch. *Left Brain, Right Brain.* New York: W. H. Freeman, 1993.

Stoddart, S., A. Bonanno, T. Gouder, C. Malone, and D. Trump. "Cult in an Island Society: Prehistoric Malta in the Tarxien Period." *Cambridge Archaeological Journal* 3, no. 1 (April 1993).

Suzuki, David, and Anita Gordon. *It's a Matter of Survival.* Toronto: Stoddart Publishing, 1990.

Thompson, Richard Frederick. *The Brain.* New York: W. H. Freeman, 1985.

Toth, Nicholas. "The Oldowan Reassessed: A Close Look at Early Stone Artifacts." *Journal of Archaeological Science* 12, no. 2 (March 1985).

Trainer, Ted. *The Global Crisis.* Sydney: University of New South Wales, 1996.

Tramo, M. J., K. Baynes, R. Fendrich, G. R. Mangun, E. A. Phelps, P. A. Reuter-Lorenz, and M. S. Gazzaniga. "Hemispheric Specialization and Interhemispheric Integration." *Epilepsy and the Corpus Callosum.* New York: Plenum Press, 1995.

Tyler, Michael J. *There's a Frog in My Stomach.* Sydney: Collins, 1984.

United Nations. *Human Development Report 1994.*

———. *The State of World Population 1994.*

———. *Global Environment Outlook 1997.*

———. *The State of World Population 1997.*

Vickers-Rich, Patricia, and Thomas Hewitt Rich. *Wildlife of Gondwana.* Sydney: William Heinemann, 1993.

Vines, Gail. "Some of Our Sperm Are Missing." *New Scientist,* 26 August 1995.

de Waal, Frans B. M. "Bonobo Sex and Society." *Scientific American,* March 1995.

———. *Good Natured.* Cambridge: Harvard University Press, 1996.

Wasser, Samuel K. "Reproductive Control in Wild Baboons Measured by Fecal Steroids." *Biology of Reproduction* 55 (1996).

Watzman, Haim. "Lusher Times at Masada." *New Scientist,* 3 September 1994.

White, Mary E. *After the Greening.* Sydney: Kangaroo Press, 1994.

———. *Listen . . . Our Land Is Crying.* Sydney: Kangaroo Press, 1997.

Wilson, Edward O. *On Human Nature.* Cambridge: Harvard University Press, 1978.

———. *The Diversity of Life.* Cambridge: Harvard University Press, 1992.

Wrangham, Richard, and Dale Peterson. *Demonic Males.* London: Bloomsbury Publishing, 1996.

Wrangham, Richard W., W. C. McGrew, Frans B. M. de Waal, and Paul G. Heltne, eds. *Chimpanzee Cultures.* Cambridge: Harvard University Press, 1994.

Zeki, Semir. "The Visual Image in Mind and Brain." *Scientific American,* September 1992.

INDEX

Aborigines
  culture, 153
  and environmental degradation, 150
  mysticism, 198
  in Tasmania, 211–13
Academy of Science in Canberra, 138
acid rain, 49–51
adolescents, 178
advertising
  agencies, 175
  and male promiscuity, 228
  and propagation, 176–77
Africa
  and ENSO events, 20–21
  extinctions, 149
  female circumcision, 229–30
  Lake Turkana, 67, 85, 202
  land degradation, 27, 30
agriculture. See also Green Revolution; land
      degradation; nutrition
  development of, 91–94
  and human plague, 236
  and mysticism, 204–5
Alaska, 23
Albert Einstein College of Medicine, 218
altruism, as genetic imperative, 173, 190
  heroism, 222–25
  parental sacrifices, 62
  and spirituality, 226–27
amygdala, 185, 186–87
Antarctica
  global warming indicators, 25
  ozone destruction, 49
  Vostok research station, 21–22, 24
anthropocentrism, 7–9
aquaculture, 43–44
arable land. See agriculture; land degradation
Aral Sea, 34
archaebacteria, 23, 110, 123, 243, 251
Asia
  acid rain in, 51
  desertification in, 27
astrology, 197
atmosphere. See global warming
attention-deficit hyperactivity disorder
      (ADHD), 170–71

Aum Shinrikyo, 253–54
Australia
  archaic humans, 81–90
  extinctions, 127, 144–47, 148–50
  gun legislation, 246
  Hamersley Range, 108–9
  land degradation, 27, 28, 30–31, 32, 36
  marine harvest decline, 43
  mysticism, 204
  North Pole, 109, 112
  nutrition, 44
  ozone destruction, 49
  Shark Bay stromatolites, 112–14
  water usage, 46
automobiles, production of, 107

bacteria. See also specific types
  reproductive process, 251–52
  waste materials, 110–14
Balbina Reservoir, 248–49
behavior. See genetics, and behavior
belief. See faith; mysticism
Bergamini, David, 225
Berlin, 208–9
Berndt, R. M., 203
Betzig, Laura, 188
Binti Jua, 62, 63
biodiversity, 15–18, 255–56. See also
      extinctions
biosphere. See also extinctions; specific
      environmental problems
  as chaotic system, 235–36
  Gaia theory, 119–20
  methane, 122–26
  photosynthetic organisms, 120–22
  stability of, 126–27
biotechnology, 39–41
Black Death, 99–101
bonobos, 134–37, 214
Bosch, Carl, 38
Bouchard, Thomas, 170, 171
bowerbirds, 179
brain
  Homo erectus, 72–81, 202
  Homo habilis, 67–69, 202
  human (see also Broca's area)

brain (*con't*)
Broca's area, 3–9, 70–72
chemical relationships in, 217–22
hemispheres of, 181–83
and mysticism, 184–87, 257–58
neurons, 165–66
plasticity of, 69–70
Broca, Paul, 4
Broca's area, 3–9, 70–72
Broecker, Wallace S., 26
Brown, Lester, 36
Butifos, 34
Butler, Juliet, 254
butterfly effect, 215–17, 234–36

Canada
hydroelectricity, 249–50
runaway warming, 23
sulfur dioxide emissions, 107
toxaphene, 48
carbon
atmospheric, 19–20, 120–22
role of, 126
carbon dioxide
and global warming, 19–20, 22, 31
and hydroelectricity, 248–49, 250
and photosynthesis, 121, 126
carbon-pollen horizons, 84–86
cars, production of, 107
Carson, Rachel, *Silent Spring*, 133
catalytic converters, 107
Cedar Lake reservoir, 250
cement production, 250
Center for Cognitive Neuroscience, 182
Centre for Energy and Environment, 45
Chain, Ernst, 98
Changeux, Jean-Pierre, 70
chaotic systems, 215–17, 234–36
chastity, 189
Chief Seattle, 203
chimpanzees, 4–7
encephalization quotient, 68
genocide, 211, 214
Moro reflex, 169
China
acid rain, 51
energy consumption, 248
land degradation, 27, 30, 32
Chi-Sang Poon, 235
chlorofluorocarbons (CFCs), 48, 120
circumcision, female, 229–30
civilization, 196. *See also* culture
cleavage sites, 216
climate, global. *See* global warming
coal, 51
codons, 59, 60
Cold War, 105
commerce. *See also* consumerism
development of, 93–94
and human plague, 255, 256
and mysticism, 127–28
Commonwealth, Scientific and Industrial Research Organisation (CSIRO), 46
communication. *See* Broca's area; language
computers, 106, 207–8

conspiracy myths, 208–9
consumerism, 12. *See also* commerce
in India and China, 248
from 1950 to 1990, 52
short-sightedness of, 178
and waste, 106
Cook, James, 140
corpus callosum, 181, 182, 183
creation-evolution debate, 129
crops. *See* agriculture; land degradation; nutrition
cuckoo, 167
cults, 253–54
culture
apocalyptic, 139–43
development of, 91–96
evolution versus, 102–9
and genetics, 152–55, 173
of hunter-gatherers, 81, 194–96
marriage rituals, 189–90
and mysticism, 187–88, 191
and polygyny, 188–89
cyanobacteria, 118

dams, 248–50
Dartmouth College, Center for Cognitive Neuroscience, 182
Darwin, Charles
Expression of the Emotions in Man and Animals, The, 220
and genetic determinism, 174
opposition to, 128, 129
Origin of the Species, The, 7
supporters, 129
Dawkins, Richard, 60–61
deceit, 178
decision-making process, 79–80
deforestation. *See* tropical forests
deoxyribonucleic acid. *See* DNA
Descartes, René, 215
desertification. *See* soil erosion
Deutsch, Georg, 182–83
de Waal, Frans, 208, 213
Diamond, Jared, 195
diet, 73, 74, 93, 195. *See also* nutrition
Dinopidae, 160
diseases, 39, 40, 251–52
Divine Wind, 223
DNA. *See also* genetics
autonomy of, 173–74
human, 6, 59, 63–65, 179–80
replication, 59–62, 103, 152
Dörner, Gunter, 132
drugs, illicit, 207
dryland salinization, 27, 31–35

Easter Island, 139–41
economics. *See* commerce; consumerism
eggs, 60
Ehrlich, Paul, 38, 52, 96, 104
Ekman, Paul, 221
El Niño, 20–21, 25, 36
emotions
and facial expressions, 219–22
and hypothalamus, 184–85

and neurotransmitters, 216–17
encephalization quotient (EQ), 68, 73, 202
endocrine disrupters, 133, 134
energy
    and agricultural productivity, 39
    consumption of, 12, 31, 52, 106, 247
    hydroelectricity, 248–50
    solar, 11, 19, 44
Engels, Friedrich, 129
ENSO (El Niño–Southern Oscillation) events,
    20–21, 25, 36
environment, instability of, 150–55. *See also* bio-
    sphere; *specific environmental problems*
ER 1470, 67, 68, 70
Erwin, Douglas H., 17
Escherichia coli, 138
estrogen, 132, 133, 217
eubacteria, 110, 123, 251
eukaryotes, 110–11
Europe
    acid rain, 50–51
    Black Death, 99–101
    extinctions, 149
    nutrition, 44
    settlers, in Australia, 150, 153
evolution, culture versus, 102–9. *See also*
    genetics
evolution-creation debate, 129
Expression of the Emotions in Man and Animals,
    The (Darwin), 220
extinctions, 15, 16, 242–43
    in Australia, 148–50
    gastric-brooding frogs, 144–47
    Maori bird massacre, 147–48
    mass, 16–18
Eysenck, Hans, 171

facial expressions, 219–22
faith, 186, 198–206. *See also* mysticism
false-memory syndrome, 182, 219
farming. *See* agriculture; land degradation;
    nutrition
Fearnside, Philip, 248, 249
feelings. *See* emotions
female circumcision, 229–30
fertility, 98, 130–31, 132–34, 217. *See also* sexu-
    ality
fertilizers, nitrogenous, 38–39
fire, domestication of, 85, 89, 201–2
Fisher, Helen, 217, 218
fish farming, 43–44
Flannery, Tim, 149
Fleming, Alexander, 98
Florey, Howard, 98
flowers, 176
Food and Agriculture Organization (FAO),
    15, 40, 41, 44
food production. *See* agriculture; land degra-
    dation; nutrition
forests, 15–16, 50. *See also* tropical forests
fraternal twins, 169, 170, 171
freshwater, 46
Friedman, Milton, 255
Friesen, Wallace, 221
frogs, gastric brooding, 144–47

fungicides, 40

Gaia theory, 119–20
gambling, 206
gas hydrates, 23
gastric-brooding frogs, 144–47
Gazzaniga, Michael S., 182
gender conflicts, 227–32
general adaptive syndrome (GAS), 131–32,
    244–45, 256
genes. *See* DNA; genetics
genetic determinism, 174
genetics, 57–59. *See also* DNA
    and ADHD, 170–71
    and chaotic systems, 216
    and human behavior
        ADHD, 170–71
        advertising, 174–75, 176–77
        brain hemispheres, 181–83
        DNA, autonomy of, 173–74
        identical twins, 169–70
        internal conflict, 175–76
        mysticism, 172–73
        rational thought, 180–81
        reflexes, 167–69
        seeing process, 171–72
        short-sightedness, 177–79
        stability, maintaining, 152
    and human origins, 65–73
    parental sacrifices, 62–63
"Genie," 70
genocide, 210–17
genomes, 59, 60
Germany, 132
global warming, 18–20, 242
    ENSO events, 20–21
    indicators of, 24–26
    polar ice caps, 21–22
    runaway warming, 23–24
Gombe Stream Research Centre, 211
Good Natured (de Waal), 213
Goodall, Jane, 211
gorillas
    language capabilities, 5–6
    encephalization quotient, 68
    infanticide, 213, 214
Gould, Stephen Jay, 102, 103
Grand Banks, 41
Great Artesian Basin, 30–31
Green Revolution, 38, 51–52, 104
greenhouse gases, 19. *See also* carbon dioxide;
    methane
gun legislation, 246

Haber, Fritz, 38
hackers, 208
Hamersley Range, 108–9
Hawaiian Islands, 49
Hawkes, Kristen, 195
heroism, 222–25, 247
high-yield crops. *See* agriculture; Green Rev-
    olution
Hitler, Adolf, 129, 189, 210
Hodges, William, 140
Holdren, John, 52

Holloway, Ralph, 67
homicide rate, 208–9
Homo erectus, 72–73
    in Australia, 81–90
    brain development in, 73–79
    decision-making capabilities, 79–80
    encephalization quotient, 202
    and language debate, 80–81
    migration of, 80
    and mysticism, 81
Homo habilis, 67–69, 202
homophobia, 189
Homo sapiens. *See* humans
homosexuality, 132, 134, 189
hormones, 130–31, 132–34, 217
horses, 64–65
Houghton, Richard, 125
humans. *See also* genetics, and human behavior
    evolution of, 240–42
    expectations of, 11–15
    origins of, 65–73
    plague indicators, 97–101
        behavior and environment, 236, 242–52
        cultural transformation, 103–5
        exponential growth, 138–43, 236–37
        fertility, 131–34, 137–38
        mysticism, 252–59 (*see also* mysticism)
hunter-gatherers. *See also* Aborigines; *Homo erectus*
    belief systems, 202–3
    Maoris, 147–48
    modern, 194–96
    territoriality, 198
Huxley, Julian, 38, 96, 104
Huxley, T. H., 129
hydroelectricity, 248–50
hydrogen, 110, 123
hyperthermophiles, 123
hypothalamus, 184–86

I = PAT, 134, 147, 242
ice ages, 72, 121–22, 125
identical twins, 169–70, 171
imagination, 90
Imo, 94
India
    energy consumption, 248
    land degradation, 30, 32
industrial pollutants, 47
infanticide, 213, 214
infatuation, 217–18
International Rice Research Institute, 45
Iraq, 80
Ireland, famine in, 40
irrigation, 32–35

jacana, 231
Japan
    hydroelectricity, 250
    kamikaze pilots, 222–25
    marine harvest decline, 43
    pollution, 51
Jenner, Edward, 97

Kahama group, 211

Kakadu Man (Neidjie, Davis, and Fox), 203
kamikaze pilots, 222–25
kangaroos, 130
Kanzi, 9, 159
Kasakela group, 211
KNM-ER 1470, 67, 68, 70
Komodo dragon, 86
Kublai Khan, 223

Lake George, 84, 85, 87
Lake Mungo, 87
Lake Turkana, 67, 85, 202
Lamarck, Jean Baptiste, 103
land degradation, 26–28, 127
    agricultural limits, 36–37
    and biotechnology, 39–41
    energy debt, 39
    Green Revolution, 38
    nitrogenous fertilizers, 38–39
    salinization, 31–35
    soil erosion, 29–31
    soil exhaustion, 29
    waterlogging, 35–36
language, 4–9. *See also* Broca's area
    and *Homo erectus*, 83–84
    learning, 69–72
    and mysticism, 196, 199–200, 257
    and rational thought, 214–15
Leakey, Richard, 16
learning, transmission of, 94
lesbianism, 134
Lewin, Roger, 16
Lincoln, Abraham, 189
Linnaeus, Carolus, 7
logic. *See* rational thought
Lorenz, Edward N., 215–16, 234–35
Lovelock, James, 119, 120, 126–27, 256
Lütken, Otto Diederich, 95–96, 150, 191

malnutrition, 44, 45
Maltese Islands, 141–43
Malthus, Thomas Robert, 94–95, 104, 150, 191
mangroves, 43–44
Maoris, 147–48
Margulis, Lynn, 118, 120, 251
marine harvest, decline in, 41–44
marriage, 189–90
Martin, Claude, 16
Marx, Karl, 95, 129
McNamara, Robert, 45
McVeigh, Timothy, 208
Meacham, Steve, 247
medicines, 147
memory, 218–19
methane
    and global warming, 19, 23–24, 126
    and hydroelectricity, 248–50
    production of, 122–24, 243–44
methanogens, 123, 124
Meyer, Wayne, 46
Middle East
    female circumcisions, 229–30
    land degradation, 27, 32, 105
militia groups, 208

Milton, Katherine, 73
mitochondria, 111
Mitumba group, 211
moa, 148
Mongolia, 27
monozygotal twins, 169–70, 171
morality, 173, 175, 183. *See also* mysticism
Morewood, Mike, 85
Moro reflex, 169
Moscow, witchcraft in, 254
Moulay Ismail the Bloodthirsty of Morocco, 189
Mungo people, 87
murder rate, 209–10
mysticism, 183–91, 231
    and altruism, 226, 227 (*see also* altruism)
    ancient remains of, 81
    in apocalyptic cultures, 140, 141–42, 143
    and commerce, 127–28
    and cultural belief systems, 196–98
    and faith, 198–206
    genetic links to, 172–73, 182
    and human plague, 252–59
    of hunter-gatherers, 194–96
    modern, 206–10

native Americans, 202–3
Neidjie, Bill, *Kakadu Man*, 203
neotony, 65, 76–77
net caster spiders, 160–65, 167
New Zealand
    extinctions, 148, 149
    and ozone depletion, 49
nitrogenous fertilizers, 38–39
nitrogen oxides, 39, 48, 50
nucleotide bases, 58
nutrition, 44–46. *See also* diet

orangutans
    DNA, 6–7
    language capabilities, 5–6
    rape, 214
Origin of the Species, The (Darwin), 7
overfishing, 41, 43
overpopulation. *See* humans, plague indicators
Owen, Richard, 129
oxygen
    and global warming, 19, 21
    bacterial, 110, 111
ozone layer, destruction of, 48–49

Pace, Norman, 123
paedomorphosis, 65
paper, consumption of, 106–7
paranoia, 208
pardalote, 193–94
parental sacrifices, 62–63
Pasteur, Louis, 97–98
peat bogs, 20, 250
pedophilia, 134
peer pressure, 188
permafrost, 23, 124
Péron, François, 211–12
pesticides, 34, 40, 47–48
pests, 39, 40
photosynthesis, 110, 112, 120–22

plague mammals, 96–101. *See also* humans,
    plague indicators
    exponential growth, 138–43
    fertility and sexuality, 130–38
Poland, 50
polar ice caps, 21–22
politics, 255, 256
pollen, 84–86
pollution, 47–48, 178
    acid rain, 49–51
    ozone destruction, 48–49
polygyny, 188–89
Polynesian cultures, 188–89
Poon, Chi-Sang, 235
population growth. *See* humans, plague indi-
    cators
Port Arthur (Tasmania), 246–47
potato blight, 40
poverty, 45
predators, natural, 39, 40
prefrontal cortex, 77–78, 79
primal myth, 209–10
promiscuity, 227, 228
promoter segment, 59
proteins, 60, 63, 216

rape, 214
rational thought, 186
    and hominids, 180–81
    and mysticism, 190–92
Raup, David M., 17
reason. *See* rational thought
reflex behaviors, 167–69
religious beliefs. *See* mysticism
replication, 59–62, 103, 152
reproduction, gastric brooding, 144–47
retiarius spiders, 160–65, 167
Rheobatrachus silus, 145–46
Rheobatrachus vitellinus, 145, 146
right-handedness, 68–69
Rome, 209
Rose River song cycle, 203
Rudd, John, 249, 250
runaway warming, 22, 23–24
Russia. *See also* Soviet Union
    land degradation, 27
    marine harvest depletion, 43
    sulfur dioxide emissions, 107
    witchcraft in, 254
Rutgers University, 217

Sagan, Carl, 184
Sagan, Dorion, 118, 120, 251
salinization, 27, 31–35
Samoan culture, 188–89
Sandeman, David, 166, 167
Savage-Rumbaugh, Sue, 9
sea levels, 83, 85–86, 244
Seattle (Suquamish Chief), 203
Sedgwick, Adam, 129
seeds, presterilized, 40–41
seeing process, 171–72
Selye, Hans, 131, 256
sex cells, 60, 133, 134
sexuality. *See also* fertility

sexuality (*con't*)
  bonobo, 134–37
  and gender conflicts, 227–32
  homosexuality, 132, 134, 189
  polygyny, 188–89
Shark Bay, 112–14
Shatz, Carla J., 69
Short, Roger, 204
Shunning (China), 51
Siberia, 244
Silent Spring (Carson), 133
Simon, Julian, 255
Sixth Extinction: Biodiversity and Its Survival,
    The (Leakey and Lewin), 16
smell, sense of, 77–79
Smil, Vaclav, 38
smoking, 178
social stress, 131, 132–33
society. *See* culture
soil erosion, 27, 29–31, 105, 150
soil exhaustion, 27, 29, 105
solar energy, 11, 19, 44
South Africa, 107
Southern Oscillation, 20–21, 25, 36
Soviet Union. *See also* Russia
  dismantling of, 103
  land degradation, 28, 34
speech, 71. *See also* Broca's area; language
sperm, 60, 133, 134
Sperry, Roger, 181
spiders
  brain of, 166–67
  retarius, 160–65, 167
spirituality. *See* mysticism
Springer, Sally, 182–83
stotting, 206–7
stress, 131, 132–33
stromatolites, 112–14
suicide, mass, 253
sulfur, 112
sulfur dioxide, 49–51, 107
Sumeria, 35
supernatural beliefs. *See* faith; mysticism

Tasmania
  genocide in, 211–13
  and gun legislation, 246–47
Tasmanian devil, 86
temperature, global. *See* global warming
temporal lobe epilepsy (TLE), 186–87
territoriality, 137–38, 194
testosterone, 132, 133, 217
Thailand, 44
thermokarsts, 23
Thorne, Alan, 83
Tokyo, 209
toolmaking animals, 159–66
Tourneur, Jacques, 198
toxaphene, 47, 48
Toyoda, Admiral, 224
transcription, 59, 60
tropical forests, 15–16, 26, 147
Truganini, 212

tundra regions, 23, 124, 126, 244
twins, 169–70, 171
Tyler, Mike, 145–46

Udayama, 188
Uganda, 40
ultraviolet radiation, 48, 49
United Nations
  acid rain study, 50
  Food and Agriculture Organization
      (FAO), 15, 40, 41, 44
  global food production study, 44
  soil erosion study, 30
United States
  acid rain, 50, 51
  ENSO events, 20
  gun lobbies, 246
  land degradation, 27, 30
  nitrous oxide emissions, 107
  nutrition, 44
  and ozone depletion, 49
  runaway warming, 23
  water usage, 46
University of Exeter, Centre for Energy and
      Environment, 45
University of Minnesota, 170
Uzbekistan, 34

vancomycin-resistant enterococcus (VRE), 251
Vernadsky, Vladimir, 119–20
"Victor," 70
violence, 213, 214. *See also* genocide
viruses, 40
vision, 171–72
Vostok research station, 21–22, 24

Wallace, Alfred Russell, 128
Washington, D.C., 208
Wasser, Sam, 130
waste materials. *See also* pollution
  human, 118
  bacterial, 110–14
water, consumption of, 46
waterlogging, 27, 35–36
Wilberforce, Samuel, 129
Wilson, Edward O., 14, 15, 18
witchcraft, 254
women
  gender conflicts, 227–32
  genetic imbalances in, 177–78
  in hunter-gatherer societies, 195
Woodwell, George, 125
World Bank, 45
World Resources Institute, 15
World War II, 222–25
World Wide Fund for Nature (WWF), 16

X-factor, 90, 252

yellow baboons, 130

Zeki, Semir, 172
zygote, 169